PESTICIDES:
FOOD AND ENVIRONMENTAL
IMPLICATIONS

PROCEEDINGS SERIES

PESTICIDES: FOOD AND ENVIRONMENTAL IMPLICATIONS

PROCEEDINGS OF AN INTERNATIONAL SYMPOSIUM
ON CHANGING PERSPECTIVES IN AGROCHEMICALS:
ISOTOPIC TECHNIQUES FOR THE STUDY
OF FOOD AND ENVIRONMENTAL IMPLICATIONS
JOINTLY ORGANIZED BY THE
INTERNATIONAL ATOMIC ENERGY AGENCY
AND THE
FOOD AND AGRICULTURE ORGANIZATION
OF THE UNITED NATIONS
AND HELD IN NEUHERBERG, 24–27 NOVEMBER 1987

INTERNATIONAL ATOMIC ENERGY AGENCY
VIENNA, 1988

PESTICIDES: FOOD AND ENVIRONMENTAL IMPLICATIONS
IAEA, VIENNA, 1988
STI/PUB/764
ISBN 92-0-010288-3

© IAEA, 1988

Permission to reproduce or translate the information contained in this publication may be obtained by writing to the International Atomic Energy Agency, Wagramerstrasse 5, P.O. Box 100, A-1400 Vienna, Austria.

Printed by the IAEA in Austria
April 1988

FOREWORD

Pesticides are an integral part of modern agriculture, also in most developing countries. Although the annual average consumption of active ingredients in agriculture may be below 0.1 kg a.i./ha, most countries now consume more than 2 kg a.i./ha; some of the intensively cropped regions in South-East Asia are exposed to even higher amounts. Inherent contamination of the environment follows if rules and regulations are not strictly adhered to.

High concentrations of pesticide residues appear worldwide, sometimes hampering trade and even occasionally posing a threat to human health. Without doubt, faunal wildlife is threatened in agricultural regions, causing economic losses such as decreased fish catch in paddy rice fields and water courses.

The search for safer, less persistent and more specific pesticides and examination of the fate of applied pesticides in various regions of the world were the main themes of the FAO/IAEA International Symposium on Changing Perspectives in Agrochemicals: Isotopic Techniques for the Study of Food and Environmental Implications. Special emphasis was placed on the use of nuclear techniques, especially on labelled compounds in research. The scientific contributions presented prove the excellence of this sometimes indispensable tool.

The symposium was jointly organized by the International Atomic Energy Agency and the Food and Agriculture Organization of the United Nations in co-operation with the Gesellschaft für Strahlen- und Umweltforschung mbH, Munich, Federal Republic of Germany. It was held in Neuherberg from 24 to 27 November 1987. The symposium commenced with two sessions that were combined with the 10th Biennial International Symposium on Chemical and Toxicological Aspects of Environmental Quality organized by the Gesellschaft für Strahlen- und Umweltforschung, the International Academy of Environmental Safety and the International Society of Ecotoxicology and Environmental Safety. The symposia were attended by approximately 200 participants from 55 countries on the first joint day, and about 60 participants from 26 Member States as of the second day.

The 2nd FAO/IAEA/GSF Research Co-ordination Meeting on Research to Develop and Evaluate Controlled Release Formulations of Pesticides to Reduce Residues and Increase Efficacy Using Radioisotopes was held immediately after and in conjunction with the symposium. The Proceedings include all the papers and posters that were presented, as well as a summary of the Research Co-ordination Meeting.

The IAEA and FAO wish to express their thanks to the Gesellschaft für Strahlen- und Umweltforschung for the generous hospitality and efficient services provided. These contributed greatly to the smooth running and success of the symposium.

EDITORIAL NOTE

The Proceedings have been edited by the editorial staff of the IAEA to the extent considered necessary for the reader's assistance. The views expressed remain, however, the responsibility of the named authors or participants. In addition, the views are not necessarily those of the governments of the nominating Member States or of the nominating organizations.

The use of particular designations of countries or territories does not imply any judgement by the publisher, the IAEA, as to the legal status of such countries or territories, of their authorities and institutions or of the delimitation of their boundaries.

The mention of names of specific companies or products (whether or not indicated as registered) does not imply any intention to infringe proprietary rights, nor should it be construed as an endorsement or recommendation on the part of the IAEA.

The authors are responsible for having obtained the necessary permission for the IAEA to reproduce, translate or use material from sources already protected by copyrights.

CONTENTS

Global pest management in the future (IAEA-SM-297/38) 1
P. Kraus
Effects of pesticides on fauna and flora (IAEA-SM-297/40) 11
P. Müller
Chemical and radioactive residues in soil: A global perspective
(IAEA-SM-297/42) 29
F.P.W. Winteringham
Pesticide dissipation in soils as a model for xenobiotic behaviour
(IAEA-SM-297/45) 45
J.B. Weber
Movement of pesticides from the site of application (IAEA-SM-297/35) 61
J.R. Plimmer
Empleo de plaguicidas y experiencias con técnicas radisotópicas en un país
en vías de desarrollo (IAEA-SM-297/14) 79
J. Espinosa-González, F. Ramón, E. Borrero de Saiz, M. Navarro,
D. Rovira, A. Guerra
Effects of ^{35}S-dimehypo pesticide on the agricultural environment and
ecosystem (IAEA-SM-297/3) 89
Z.Y. Chen, C.Y. Mi, D.C. Ye, B. Chen
Losses of pesticides from agriculture (IAEA-SM-297/21) 101
J.K. Kreuger, N. Brink
Controlled release insecticide formulations for tropical application
(IAEA-SM-297/27) 113
L. Vollner, A. Ghods-Esphahani
El uso de plaguicidas en América Latina: Tendencias e implicaciones
ambientales (IAEA-SM-297/36) 123
M.A. Constenla
Distribution of residues of ^{14}C-aldicarb applied to cotton plants in
Gezira, Sudan (IAEA-SM-297/20) 149
G.A. El Zorgani, T.N. Bakhiet, N. Sharaf Eldin
Fate and magnitude of malathion residues in stored wheat and barley
(IAEA-SM-297/23) 157
K. Gözek, F. Artiran
Efficacy of the controlled release of ^{14}C-carbofuran formulation for
pest control in cotton (IAEA-SM-297/31) 169
F.F. Jamil, M.J. Qureshi, A. Haq, N. Bashir, S.H.M. Naqvi
Fate of pesticides in soils of various organic matter (IAEA-SM-297/10) 177
L. Horváth, F. Kling, L.P. Simon

Field evaluation of controlled release carbofuran formulations in
Indonesian flooded rice (IAEA-SM-297/34) 185
R.M. Wilkins, Haeruddin Taslim, Hendarsih Suharto
Development and testing of pesticide formulations with thermoplastic
polymers (IAEA-SM-297/32) 195
F. Korte, G. Pfister, M. Bahadir
Seed dressing with controlled release formulations: Evaluation using a
radioisotope technique and yield estimations for the control of aphids
and stem nematodes in field beans (*Vicia faba* L.) (IAEA-SM-297/2) 205
B.C. Schiffers, P. Dreze, J. Fraselle, M.C. Gasia
Current trends in pesticide usage in some Asian countries:
Environmental implications and research needs (IAEA-SM-297/30) 219
M. Soerjani
Fate of ^{14}C-carbofuran in model rice/fish and rice/fish/*Azolla* ecosystems
(IAEA-SM-297/4) 235
Jinhe Sun, Jianying Gan, Yongxi Zhang
Studies on the effects of some insecticides on the brain acetylcholinesterase
activity of *Tilapia zilli* in two treated tropical rivers (IAEA-SM-297/8) ... 247
L.A.K. Antwi
Distribution and metabolism of carbofuran in paddy rice from
controlled release formulations (IAEA-SM-297/28) 257
L. Vollner, R. Kutscher, J. Dombovári, M. Oncsik
Improved alginate based slow release pesticide formulations
(IAEA-SM-297/33) 267
E. Schacht, J.C. Vandichel
Application of *Bacillus sphaericus* in the control of *Culex fatigans*
(IAEA-SM-297/12) 277
S.V. Amonkar, A.S. Rao, V. Narayanan

Poster Presentations

Impacto de plaguicidas en zonas agrícolas de los Llanos Orientales
(Colombia) (IAEA-SM-297/26P) 289
J.A. Avellaneda
Residuos, sorción y desorción del herbicida paraquat en suelos tropicales
(IAEA-SM-297/5P) 290
M.A. Constenla, L.E. Mora, E. Rojas, E. Carazo
Residuos de los herbicidas diurón y ametrina en tres suelos cubanos
(IAEA-SM-297/6P) 292
M.M. Hernández-Alfonso, A. Travieso

Effects of slow release formulations of herbicides on aquatic vegetation
(IAEA-SM-297/39P) 295
M. Soerjani

Изучение поведения и контроль персистентных пестицидов в почвах
сельхозугодий (IAEA-SM-297/13P) 297
М. И. Лунев
*(Study of the behaviour and control of persistent pesticides in
agricultural soils: M.I. Lunev)*

Distribution of ^{14}C-tridemorph after application as soil drench
(IAEA-SM-297/25P) 300
R.B. Mohamad, T.K. Lim, R.T. Hamm

Radioisotopic studies on the content and conversion rates of toxic
impurities in malathion formulations (IAEA-SM-297/17P) 301
W. Reimschüssel, L. Adamus, B. Śledziński

Biodegradation of aldrin and gamma-BHC in tropical soils:
Control of *Callosobruchus maculatus* F. (IAEA-SM-297/19P) 303
H.G. Morgan

Influence of placements on the distribution of controlled release
^{14}C-carbofuran in water and rice plants in Thailand
(IAEA-SM-297/29P) 306
Paiboon Prabuddham, Nuansri Tayaputch, Bandhit Anuruk

Pesticide residues in Thailand (IAEA-SM-297/22P) 308
Nuansri Tayaputch

Summary Report of the Research Co-ordination Meeting 313

List of Chairmen of Sessions and Secretariat of the Symposium 319
List of Participants 321
Author Index 329
Index of Papers and Posters by Number 331

Invited Paper

GLOBAL PEST MANAGEMENT IN THE FUTURE

P. KRAUS
Biologische Forschung,
Pflanzenschutzzentrum Monheim,
Bayer AG, Leverkusen,
Federal Republic of Germany

Abstract

GLOBAL PEST MANAGEMENT IN THE FUTURE.
 In spite of the problems that have arisen with overproduction in parts of the northern hemisphere, agriculture has to meet the challenge of providing enough food within the next 50 years for as many people as have lived on the earth during the last 1987 years. With a constantly growing world population this can only be achieved by intensifying production on a given acreage, since there are only marginal possibilities for extending agricultural land. Therefore, the yield potential of crop plants has to be secured against detrimental effects. The input for pest and weed control amounts to an average of less than 5% of the running costs at farm level but stabilizes the yield and quality of the agricultural production significantly. Rational use of biological and cultural practices in addition to chemicals will grow in importance to keep losses below the economic threshold. These approaches to pest management have global dimensions; intensity and structure, however, vary from region to region. The highly intensive agriculture of developed countries is increasingly subject to aspects of agricultural and environmental policy, whereas political economy and food supply are the dominating influences in developing countries and have a direct impact on pest management methods. These more regional trends do not always run parallel with the international policy trends of the agrochemical industry. However, research for effective crop protection basically envisages common targets; some aim at low rates, selectivity and favourable toxicological and ecotoxicological properties. Regional conditions may therefore be covered by efforts to select suitable products, formulations and application techniques. This will only be possible through an integrated research strategy, including biochemistry and molecular biology, based upon future oriented market analysis.

1. INTRODUCTION

With the tremendous growth in the world's population, major challenges will have to be faced in the coming decades. Concerning food supply, the agricultural self-sufficiency of a single country is not the main issue, rather the global food supply.

Problems with overproduction in some countries and famine in others, difficulties with the transfer of technology between them and growing ecological concerns in a world with limited resources sometimes combine to encourage scepticism towards research and doubts as to whether science will be able to solve future problems [1]. It would be dangerous if this trend should lead to a reduction in the efforts being made to assure a better food supply in the future. Pest management, certainly, is an important part of these efforts.

2. CROP PROTECTION AND PEST MANAGEMENT

Pest and weed control is not an end in itself but a way of protecting the crop plant, thereby ensuring its yield and quality. However, complete eradication of diseases, pests or competing weeds is not possible and, apart from the ecological consequences of such drastic action, is also not necessary because to a certain extent the crop plant can depend on its own competitiveness and defence mechanisms. Only when damage surpasses economic thresholds do crop protection measures become necessary.

Pest management means the sum of all efforts to keep pests, diseases and weeds below this damage level with suitable products or methods, using efficient application techniques and the right timing, with the necessary follow-up, and by maintaining a balance between the economy and the ecology. The success of such pest management will, in part, depend on whether plant production can keep pace with the growing world population.

3. WORLD POPULATION AND FOOD SUPPLY — THE NEED FOR CROP PROTECTION

The current population density in the Federal Republic of Germany is 250 people/km^2, whereas 160 years ago in the same region it was 50 people/km^2 (see Fig. 1). At the same time, wheat yields rose from 1 to 5 t/ha. The fivefold increase was possible through a combination of plant breeding, fertilization, mechanization and crop protection, proving that by mobilizing all these efforts the production potential still shows some promise.

With the same factor of 5, world population has grown from 1 billion in 1820 to 5 billion in 1987.[1] Because the growth shows no sign of abating, calculations of the food supply cannot be extended beyond the year 2000 [2].

---

[1] 1 billion = 10^9.

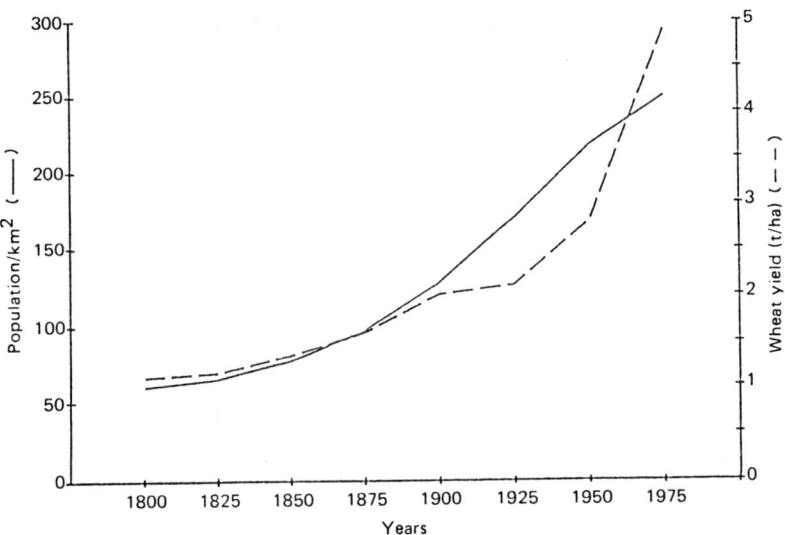

FIG. 1. Population per square kilometre and wheat yields in the region now known as the Federal Republic of Germany (1800–1975).

From the figures of world population and total agricultural land shown in Table I, the available land per capita has decreased from 3600 m^2 in 1970 to 3000 m^2 in 1981, with an expected further reduction to 2200 m^2 by the year 2000. The question of whether such an area is sufficient for the necessary food supply can be answered by looking at the conditions in Europe, where only 2200 m^2 per capita are currently available.

The increase in plant production that is necessary at least to make up for the relative decrease in agricultural land has to materialize — according to the FAO study Agriculture: Towards 2000 [3] — mainly from higher production intensity on the given land, as seen in Table II. Intensification of agricultural production, however, is not possible without efficient protection of the crop plant. Since in developed countries the level of total production has to be secured rather than increased, the need for higher intensity of production in developing countries is an even more urgent challenge for crop production.

4. CROP PROTECTION AS A PRODUCTION FACTOR — INPUT AND EFFECT

Twenty years ago, Cramer [4] estimated that the losses in plant production up to harvest caused by pests, diseases and weeds totalled one-third of the potential yield; the detailed figures are given in Table III.

TABLE I. WORLD POPULATION AND TOTAL AVAILABLE AGRICULTURAL LAND

	1970	1981	2000
Agricultural area, world (million km^2)	13.30	13.60	14.00
Population, world (billion)[a]	3.70	4.50	6.30
Agricultural area per capita (m^2)	~3600	~3000	~2200
Agricultural area, western Europe (million km^2)		0.83	
Population, western Europe (billion)		0.37	
Agricultural area per capita (m^2)		~2200	

[a] 1 billion = 10^9.

TABLE II. AGRICULTURAL PRODUCTION RESOURCES IN DEVELOPING COUNTRIES (1975–2000) [3]

	Expansion of agricultural land (%)	Enhanced cultivation (%)	Yield increase per unit of land (%)
Africa	27	22	51
Far East	10	14	76
Near East	6	25	69
Latin America	55	14	31
90 developing countries (average)	26	14	60

TABLE III. WORLDWIDE LOSSES IN PLANT PRODUCTION [4]

	Loss in % caused by			Total
	Pests	Diseases	Weeds	
North and Central America	9.4	11.3	8.0	28.7
South America	10.0	15.2	7.8	33.0
Europe	5.1	13.1	6.8	25.0
Africa	13.0	12.9	15.7	41.6
Asia	20.7	11.3	11.3	43.3
Oceania	7.0	12.6	8.3	27.9
USSR and China	10.5	9.1	10.1	29.7
$\bar{X}$	12.3	11.8	9.7	33.8

More recent estimates [5] have indicated a similar global situation. In western Europe, with its intensive crop protection practices, the loss is sometimes much lower, mainly in those areas where, for example, under favourable conditions 10 t/ha of wheat can be harvested today. On the other hand, insufficient or inadequate crop protection in developing countries cannot always prevent a total loss.

The low intensity of current production in some developing countries suggests considerable potential for yield increases under locally appropriate conditions of pest management and agricultural practice; compare, for example, the average wheat yield of 1.53 t/ha in developing countries with that of 3.67 t/ha in western Europe.

The relative cost of pest management will, of course, depend on the economic benefit a farmer can expect for his plant production. In the Federal Republic of Germany, with its intensive agriculture and hence high input of crop protection, the cost was only 2.7% of the total agricultural production costs in 1981.

The FAO study [3] concluded that continuation and extension of crop protection measures will be necessary until the year 2000 [6].

5. INTEGRATED PEST MANAGEMENT — ROLE OF CHEMICAL CROP PROTECTION

Chemical crop protection is an indispensable component of integrated pest management, which itself is a substantial part of the system of agricultural practices called integrated crop production. Chemical crop protection, including choice of a suitable pesticide of adequate formulation and dose, correct application at the right time and observation of the economic thresholds, will not be successful unless agricultural practices, mechanical and physical practices and biological and biotechnological practices are taken into consideration. On the other hand, today there is little doubt that chemical crop protection must continue to play a major role in effective integrated pest management.

This situation is unlikely to change in the near future because 8 to 10 years are needed to market a new compound. Even projections into the next century will not result in any significant changes unless an immediate and fundamental turn is made towards new research goals.

Progress in the field of biotechnology is visible but is still largely confined to basic research or prototype testing and so far suggests little or no alternative to chemical crop protection beyond supplement or support [7].

6. REQUIREMENTS AND REALIZATION — GENERAL AND REGIONAL ASPECTS OF PEST MANAGEMENT

Chemical crop protection must be adaptable to the needs resulting from the interaction between the economy and the ecology.

Today's requirements result from the experience and success achieved so far, which can also be projected into the future. A pesticide will only be successful on the market when it provides advantages over existing compounds. Conditions for the successful development of new pesticides can be illustrated as follows:

Rate: While older pesticides were known to be active in the kilogram range, recently developed compounds are already effective in the gram range.

Formulation: Pesticides have to be formulated in such a way that the active ingredient can be distributed as evenly as possible. Formulation technology should also improve the handling and application of the pesticide. Liquid formulations without organic solvents and dry formulations with decreased dusting provide improved safety.

Activity spectrum: While in the past a farmer could use only broad active pesticides, he now has a wide choice from which he can select a specific compound or a combination product which exactly suits his situation.

Mode of action: Special properties of a compound allow greater flexibility in pest management. The systemic property, for example, provides protection even after penetration of the pest into the plant tissue. Selective herbicides can be applied even after emergence of the crop plant.

Environmental research: Only compounds which pass strict toxicological tests, show no accumulation in the soil and cause no groundwater contamination through leaching will be chosen for further development. The protection of beneficial organisms is an additional criterion.

The final results of all these research efforts are crop protection products which meet today's requirements of efficacy, safety, economy and ecology. These criteria will also have to be met by tomorrow's pesticides, whether based on conventional chemical synthesis or biotechnology. Therefore, development of a new pesticide will take 8 to 10 years and cost in excess of DM 100 million. Only major crop protection problems and world markets can justify these efforts and costs. The consequence of such an internationalization of crop protection would be a concentration of the pesticide producing industry. Only those companies which operate on a worldwide basis and have their own broad research and development facilities and the necessary financial support for long term operations will survive.

In contrast to the crop protection industry, which is becoming more and more international, pest management has no uniform global dimension. It is greatly influenced by regional aspects which lead to differing concepts of agriculture and crop protection. In developed countries emphasis is no longer placed on enhanced yield, but on stabilizing the yield and quality. Agricultural and environmental politics increasingly influence pest management concepts:

Agricultural policy: Fifteen years ago, in Europe, sunflowers were not regarded as a potential crop for intensive production and, hence, justifying crop protection. Today, the cultivated acreage is increasing, demanding special R&D efforts for compounds, new formulations and application technology.

Environmental policy: For ecological reasons some countries are striving to reduce chemical crop protection. This can only be achieved if all the other aspects of the integrated pest management are effective enough to fill the gap. Research on further aspects, from breeding for resistance to the search for biological alternatives for the chemicals, must therefore be intensified.

Contrary to the trend in developed countries, improvement in food supply is still the driving force in developing countries, with great emphasis placed on pest management concepts, e.g.:

Food supply: The Green Revolution, which aimed at improving agricultural self-sufficiency, was only successful when high yielding varieties were given the necessary fertilizer and when the farmers realized the need for adequate pest management.

Political economy: Developing countries are also able to finance their food requirements by growing cash crops for export purposes. Egypt, for example, depends on

TABLE IV. DISTRIBUTION OF R&D COSTS
FOR PESTICIDES (percentage of total)

	1975	1988
Synthesis	18.3	17.6
Biological testing	32.1	30.9
Toxicology/ecobiology	18.0	32.6
Process development	24.4	11.1
Formulation, packaging, etc.	7.2	7.8

the income from high quality cotton for its national budget, therefore successful pest management is viewed as a national effort.

As pest management programmes vary from country to country, the single farmer has the common aim of protecting his crop and income from losses.

7. MANAGEMENT OF R&D FOR CROP PROTECTION

Pest management can only be applied successfully if R&D efforts are managed accordingly. All the activities necessary for the development of new pesticides (see Table IV) have to be well co-ordinated in an elaborate time frame. The cost of environmental research (toxicology, ecobiology) has increased from less than one-fifth of the total R&D costs in 1975 to an expected one-third in 1988.

The crop protection industry regards this effort as necessary for the safety of the producer, the applicator, the consumer and the environment, but it has to be stressed that registration of the pesticides has to be managed in order to standardize and harmonize the requirements. When national or even state regulations differ so greatly that the R&D efforts have to be multiplied, fewer and fewer companies will be able to afford these expenditures [8].

Management of crop protection is evident both within and between companies. Strategies to prevent the development of pesticide resistance have been discussed by groups such as FRAC or IRAC (Fungicide or Insecticide Resistance Action Committee), with the participation of the companies involved. The companies organized within GIFAP (Groupement international des Associations nationales de fabricants de produits agrochimiques) have committed themselves to the FAO Code of Conduct, which guarantees the safety and quality of the pesticides from production to marketing and from technical service to advertising.

This type of crop protection management will have to become more of a reality in the future if the demands for high quality, healthy food in sufficient quantities and at affordable prices are to be met.

REFERENCES

[1] Our Common Future, World Commission on Environment and Development, Oxford University Press, Oxford (1987).

[2] KREMER, F.W., Moderner Pflanzenschutz — Sicherung der Ernährung der Weltbevölkerung, Forum — Fond der Chemischen Industrie **21** (1983) 7–15.

[3] Landwirtschaft 2000, Schriftenreihe des Bundesministeriums für Ernährung, Landwirtschaft und Forsten, Heft 274, Landwirtschaftsverlag, Munster (1982) (Translation of FAO Report, Agriculture: Towards 2000, Conf. Doc. C.79/24, FAO, Rome (1979)).

[4] CRAMER, H.H., Plant Protection and World Crop Production, Pflanzenschutz Nachrichten Bayer **20** (1967) 1–524.

[5] AHRENS, C., CRAMER, H.H., MOGK, M., PESCHEL, H., "Economic impact of crop losses", 10th International Congress of Plant Protection 1983 (Proc. Conf. Brighton, 1983), Vol. 1 (1984) 65–73.

[6] KRUSEN, F., Ernährung 2000: Sechs Milliarden Menschen können satt werden, Olzog, Munich (1986).

[7] KRAUS, P., "Die Biotechnologie der Pflanzenforschung", Presse-Seminar Forschungsschwerpunkte bei Bayer, Bayer AG, Leverkusen (1985) 29–39.

[8] BÜCHEL, K.H., "Political, economic and philosophical aspects of pesticide use for human welfare", IUPAC Pesticide Chemistry: Human Welfare and the Environment (MIYAMOTO, J., et al., Eds), Pergamon Press, Oxford and New York (1983).

Invited Paper

EFFECTS OF PESTICIDES ON FAUNA AND FLORA

P. MÜLLER
Institut für Biogeographie,
Universität des Saarlandes,
Saarbrücken,
Federal Republic of Germany

Abstract

EFFECTS OF PESTICIDES ON FAUNA AND FLORA.
The complexities of environmental interactions in the transport, transformation and biological effects of widespread, low level chemical contaminants confound the ability of current methods to predict the possible health and environmental impacts of both new and old chemicals. We cannot transfer with certainty the knowledge gained in the laboratory of the behaviour and effects of a single chemical study in isolated living species to that gained in a natural ecosystem. However, over the past years the environmental monitoring programmes have undergone rapid conceptual and technological advancement. Environmental monitoring plays an essential role in the evaluation and management of pesticides and other anthropogenic chemicals. In the absence of effective environmental monitoring, detection of serious environmental contamination and threats to human health caused by chemical pollutants may occur only after critical and irreversible damage has been done. Between 1976 and 1987 the ecosystematic effects caused by various herbicides and insecticides were analysed. The regenerative power of the tropical ecosystems, after having been treated with insecticides, does not only depend on the type of pesticide or its formulation and concentration, but also in a decisive way on the proportions of sprayed and unsprayed areas and the ecophysiological capacity of different organisms in the tropical food chains.

1. CHLOROHYDROCARBONS, CARBAMATES, ORGANOPHOSPHORUS COMPOUNDS AND PYRETHROIDS

Application of insecticides is currently the most common method used for controlling or eradicating pest arthropods [1]. All the insecticides used for this purpose are non-specific, i.e. they not only kill the targets, but also affect other organisms, non-vertebrates and vertebrates [1, 2]. Damage to these non-target organisms may be reflected by sublethal, physiological or behavioural effects, direct mortality, long term population reduction or, in the case of persistent insecticides, be caused by bioaccumulation within the food web [3–7]. The magnitude of the adverse effects

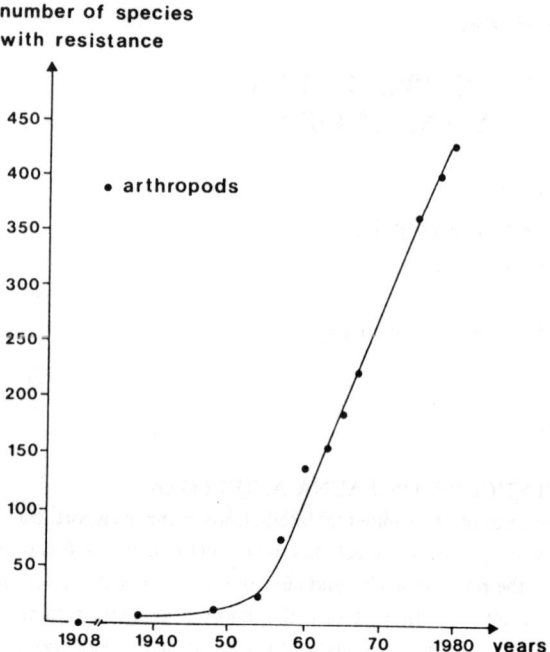

FIG. 1. *Number of arthropod species with various types of resistance [12].*

depends on a range of factors, particularly the application rate and quality [7–9], the type of insecticide and formulation, the mode of application, the type of landscape covered including the vegetative cover and presence or absence of stagnant or running water, and the composition of the food web itself. Insecticides are never target specific and aerial dispersion of insecticides over thousands of square kilometres must be regarded as potentially hazardous to man and the environment [10, 11]. The wide range of indigenous animal and plant species have all undergone the process of organic evolution and have many biochemical patterns in common. The specificity of toxic substances is thus a relative matter and it is likely that an insecticide will have some effect on various forms of life other than insects. The trend of pesticide evolution over the past 30 years, however, has been towards increased specificity. Concurrent with pesticide evolution, some arthropod species have proved to be resistant to pesticides [12]. Development of new insecticides may be considered a practical reaction to the development of resistance (Fig. 1 and Table I).

The principal insecticides utilized before World War II were inorganic compounds and/or inhibitors of the carbohydrate oxidation that produces adenosine triphosphate. The exceptions were two neurotoxic insecticides, pyrethrum and nicotine. Synthetic organochlorine insecticides were found to be remarkably residual.

TABLE I. SUCCESSION OF INSECTICIDES EMPLOYED TO CONTROL THE WATER MOSQUITO *Aedes nigromaculis*, THE BUDWORM *Heliothis virescens* ON COTTON AND THE TWO SPOTTED SPIDER MITE *Tetranychus urticae*, 1946–1976 [12]

Aedes nigromaculis	*Heliothis virescens*	*Tetranychus urticae*
DDT	DDT	Parathion
HCH, aldrin	Toxaphene, endrin	Azinphosmethyl
Parathion, malathion	Malathion	Carbophenothion
Methyl parathion	Methyl parathion	Ovex, fenson
Fenthion	Monocrotophos	Chlorobenzilate
Chlorpyrifos	Carbaryl	Dicofol
Diflubenzuron	Chlordimeform	Tetradifon
	Synthetic pyrethroids	Propargite
	Bacillus thuringiensis	Chlordimeform
	Heliothis virus	Pentac
		Oxythioquinox
		Cyhexatin

Organophosphorus compounds share a common chemical structure, but they differ greatly in the details of their structure and in their physical and pharmacological properties and, consequently, in their use or in their proposed purpose. The toxic signs and symptoms that characterize poisoning by nearly all organophosphorus compounds are thought to depend on the inhibition of acetylcholinesterase. Carbamate insecticides inhibit cholinesterase in both insects and mammals when the substitutions on nitrogen are hydrogen or methyl groups and when the substitution on ether oxygen is a relatively large moiety. The biological action of the pyrethroids depends on a disturbance to axonic nerve impulse conduction. In contrast to chlorohydrocarbons (especially DDT), organophosphates and carbamates, the precise mode of action of these pyrethroids is unknown.

TABLE II. LD$_{50}$ VALUES (ng) FOR DELTAMETHRIN AND ENDOSULFAN IN *Glossina morsitans* [14, 15]

Temperature (C°)	Deltamethrin (D)	Technical grade endosulfan (E)	Factor (E/D)
5	0.0046	7.95	1630
10	0.0082	5.98	729
12	0.0103	5.33	517
15	0.0146	4.49	308
19	0.023	3.57	155
20	0.026	3.37	130
25	0.046	2.53	55
30	0.082	1.90	23

2. ECOTOXICOLOGY AND ECOLOGICAL EFFECTS

2.1. Formulation, application and ecological effects of controlled release pesticides

The magnitude of adverse ecological effects depends on a range of factors, particularly the application rate, the type of insecticide and formulation, the mode of application, the landscape and the composition of the food web. With the active ingredient alone we are unable to predict the reaction of natural ecosystems [1, 13, 14]. Our LD$_{50}$ values are valid only for organisms, not for biocenoses. This is why it is difficult to transfer laboratory results to natural ecosystems. The dose effect determined for many singular elements can obviously vary considerably if, for instance, there is a change in temperature (Table II) [14, 15].

The values given in the table clearly show that when the effects of singular elements are analysed, the dose effect, if any is present, is only valid for a specifically defined situation. Furthermore, the following points have to be taken into consideration:

(1) The relationship will depend on the chosen type of reaction
(2) The dose effect only reflects the susceptibility of individuals and not the whole population

(3) Because of the complexity of reactions in the biosystems, only acute mortality doses can generally be observed
(4) For cancerogenic compounds, no effect levels can be defined.

Nowadays, we are aware that chemical mixtures will have different effects, i.e. they may show an additive effect (chlororganic solutions), a synergistic effect or an antagonistic reaction. In 1985 the United States Environmental Protection Agency proposed that the additivity should be assumed if there is insufficient data on the effects of the chemical compounds. However, the Maximum Working Place Concentration Commission of the Deutsche Forschungsgemeinschaft does not accept this proposal [14].

2.2. Residual spraying and non-residual sequential aerosol spraying techniques

Residual spraying with dieldrin, endosulfan, cypermethrin or deltamethrin is currently the most common method used for controlling or eradicating insects, e.g. tsetse flies in Africa [16].

Either fixed wing aircraft or helicopters may be used for spraying insecticides with the sequential aerosol application techniques. First, fixed wing aircraft may be employed to apply insecticides in the form of aerosol droplets which are widely dispersed over savanna woodlands in order to make direct contact with the flies. However, only adult flies are vulnerable to this technique and repeated applications are necessary until all the pupae have emerged; each spray cycle must be completed before the female flies deposit their larvae in the soil. Although residual insecticides of the organochlorine type can be used, this technique is mainly used for non-residual chemicals. The most common insecticide in regular use today for *Glossina* is endosulfan. Up to six applications (low doses (14 to 20 g/ha)) may be necessary, but four are generally sufficient. On the basis of the results of experiments carried out in NE Zimbabwe and Zambia in 1986 and 1987 it was found that endosulfan application by the sequential aerosol spraying technique is environmentally acceptable.

The second aerial technique involves helicopters; for example, they have been used in Cameroon (Adamaoua) to apply residual dosages of insecticides to vegetation at the edge of rivers and woodlands to control riverine and savanna flies, respectively.

2.3. Impact on targets and non-targets

2.3.1. Residual spraying in Cameroon

In Cameroon single applications of insecticides such as dieldrin or endosulfan are sprayed at dosage rates of approximately 1000 g/ha (the deltamethrin and cypermethrin doses range between 20 and 30 g/ha). In practice, a single application of

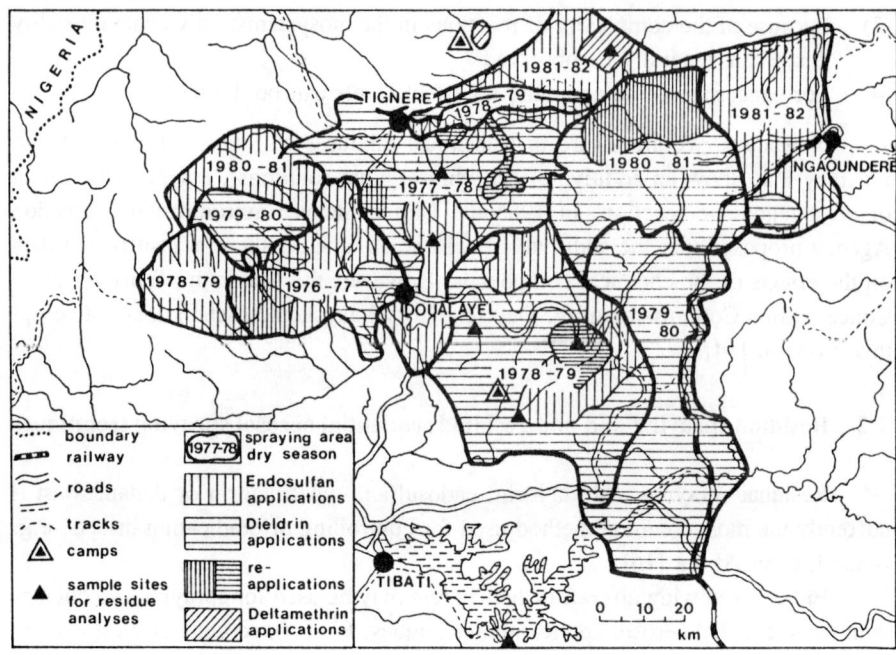

FIG. 2. *Helicopter spraying in Adamaoua (Cameroon) (up to the end of 1982).*

dieldrin or endosulfan should secure permanent control of the flies (Fig. 2). Tsetse flies have been eradicated from some areas of West Africa where this technique has been supplemented by ground and helicopter spraying. Most of the insecticides commonly used in tsetse control and tsetse eradication programmes also cause mortality in non-targets. For this reason, the ecological effects caused by various insecticides such as DDT, dieldrin, endosulfan, deltamethrin and cypermethrin were analysed between 1976 and 1987 in Cameroon. It was found that after a dieldrin application of 900 g/ha, the effects were still detectable after 3 years, as proved by the poor population density of Gryllidae, Staphylinidae, Tenebrionidae and Diptera in particular, as well as by the residues in vertebrates. The residue values were highest 6 to 12 months after application.

Regression in and destruction of some populations were observed (Chiroptera, Soricidae). However, 5 years later vital populations of these species could again be found [1]. The residue values detected in the non-migratory species have remained under the detection limit. After 7 years, considerable changes in land use had taken place, at least near Minim. In the area free of tsetse flies, the vegetation was not sufficiently abundant at the end of the rainy season to feed the cattle and their young. There were also clear indications of overgrazing. The gallery forest had been destroyed by men and cattle alike. Larger vertebrates (e.g. *Kobus kob, Alcelaphus buselaphus, Hippotragus equinus*) were rare or had completely disappeared from the area. It should be pointed out that it is not the application of chemicals but the uncontrolled land utilization that has destroyed the gallery forest ecosystems. The highest residue values can still be found in areas where ground spraying was carried out over the last 4 years. In general, it has been shown that deltamethrin, cypermethrin or endosulfan have little residual effect. For this reason, a guaranteed barrier effect can currently only be obtained by two insecticides, i.e. DDT and dieldrin, both of which are prohibited in the European Economic Community and the United States of America.

A number of endosulfan application areas have been analysed throughout Cameroon. Three years after spraying with endosulfan at a dosage of approximately 1000 g/ha no negative effects were observed on the gallery forest ecosystems situated in the study area north of Tignère. This can be attributed to the insecticide. However, various susceptible insectivorous species had returned. In particular, they dominated the avifauna. The residues had fallen below the detection limits. In the sprayed area west of Galim, where endosulfan was applied in 1987, adverse effects on the aquatic fauna were repeatedly observed. In smaller gallery forests, where small streams were more strongly exposed, high concentrations of the insecticide were detected on the water surface.

From the ecological point of view, multiple applications of insecticides are not acceptable for 3 years (endosulfan) and 5 years (dieldrin). The ecological damage observed in the region of the Mayo Dankali (related to the high residue values for dieldrin) clearly indicates that the survival limits have been reached for savanna ecosystems. Regarding their regenerative power, the negative effects of insecticides are more serious in these regions than, for example, savanna fires or overgrazing by cattle.

Deltamethrin applied at concentrations of 20 to 30 g/ha, had an acute effect on epigean arthropods. The arthropod biomass was reduced by approximately 50% in the application areas. The total aquatic fauna was also severely affected. Eight weeks after spraying, shrimps (Natantia), Hydrometridae, Dytiscidae, Trichoptera and Culicidae were still absent.

Five years after application, susceptible species again appeared in the sprayed areas north of Tignère. Because transhumance had obviously been better monitored, at least in 1987, no extensive savanna fires had occurred and many large animals were observed (*Alcelaphus buselaphus, Sylvicapra grimmia, Ourebia ourebi, Cephalophus rufilatus*, etc.). There was a preponderance of insectivorous species in the avifauna.

National parks pose a special problem. According to our analyses, gallery forest ecosystems need about 5 years to regenerate after a dieldrin dose of only 900 g/ha; 3 years are needed after an endosulfan dose of up to 1000 g/ha. Over this whole period, the residue values were measured, at least for the soil. The regenerative power depends on the size of the area sprayed (gallery forest, density of vegetation, unsprayed savanna). It follows from these findings, as well as from the

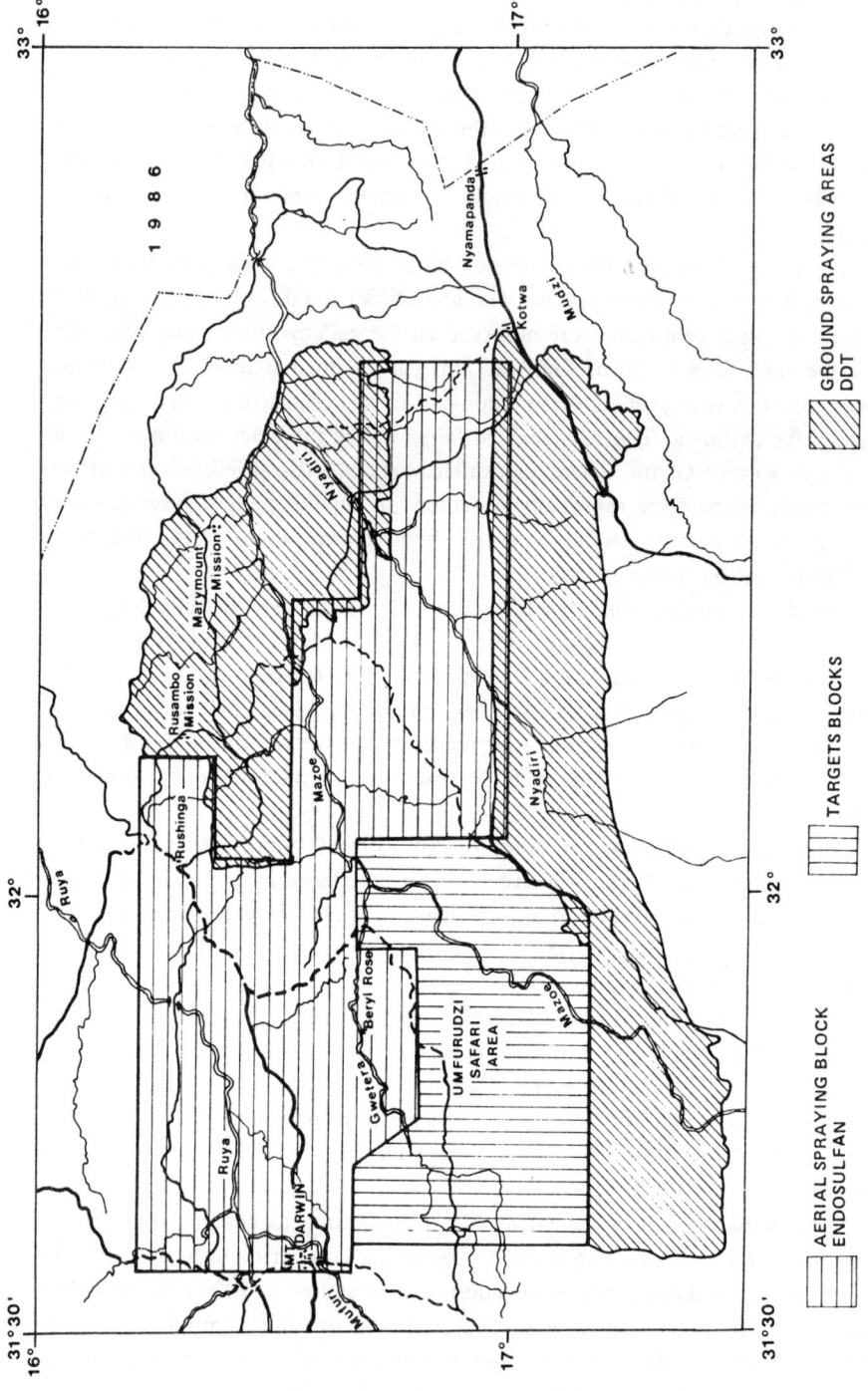

FIG. 3. *Aerial spraying block with endosulfan, target blocks (14 to 20 g/ha) and the national DDT ground spraying areas in Zimbabwe (1986).*

biocenological results, that application of insecticides in national parks is unacceptable. However, large populations of *Glossina* exist in these areas. For this reason, the population density will have to be reduced along the borders of the parks. To provide additional protection, the existing protected areas, e.g. for professional hunting or game ranging, would have to be enlarged. Our experience in Africa has shown that this form of land utilization would guarantee additional protection for the national parks. In the protected areas surrounding the national parks, tsetse populations should be controlled by odour baited targets. Residual spraying could be carried out on the border areas. In Cameroon, we have seen that every insecticide application is connected with an adverse ecological disturbance of the ecosystem and, therefore, is only acceptable if it justifies further utilization that is of advantage to the people concerned.

Unlike the apprehension sometimes raised in public discussions, we would like to point out that the insecticide doses used for tsetse control are considerably lower than other pesticides used in the tropics. However, natural ecosystems are affected more intensely by tsetse control than by agricultural utilization. Nevertheless, the most adverse effects on the gallery forests are produced by fire or extensive clearing to obtain more farmland.

2.3.2. Non-residual sequential aerosol spraying techniques in Zimbabwe

The tsetse control operation carried out in NE Zimbabwe during the 1986/1987 dry seasons consisted of one main aerial control operation. Endosulfan (14 to 20 g/ha) was applied to an area of approximately 3200 km^2 in 1986. For technical reasons this area was divided into blocks: one block with odour baited targets, partly overlapping with block 2 of the endosulfan treated area and block 3 where DDT ground spraying was carried out by the Zimbabwe Government (Fig. 3). At the SE corner of the main control area, one separate block of approximatley 500 km^2 was sprayed with deltamethrin. The timing of the aerial application of endosulfan (five spraying cycles) depended on the actual temperatures which influence the length of the pupae stage. For each cycle, fixed winged aircraft with Micron-Air equipment sprayed the insecticide from about sunset until 02:00 or 05:00 hours at about 15 m above the tree tops; the distance between each aircraft was on average 200 m.

The methods used for monitoring the effects of the endosulfan application were selected so that together they would permit evaluation of the impact of the insecticide for the ecosystemic food web as a whole. Additionally, the effects on humans were studied. Most of the investigations were conducted at the main study site and an unsprayed control area.

Condition indices showed that the health of the birds, small mammals and cattle was unaffected throughout the whole spraying period. Gonad development in birds and small mammals underlined the good condition of these animals.

Immediate effects on terrestrial insects were clearly detectable, but no long term effects were found. There was no evidence that endosulfan killed honey bees. Immediate effects on aquatic arthropods in running water were also clearly detectable, but only predatory water beetles (Laccophilinae) were severely reduced. No effects were found on emerging insects, either in stagnant or in running water.

The fish population in rivers was clearly affected, whereas in dams only a minor number of fish died. The highest fish mortality was observed after first cycle spraying, but even at the end of operations small as well as larger fish specimens were still living in the river. Even a low dosage (14 g/ha) affects the fish population. Residues of endosulfan were detected in only 3% of all the samples taken (except fish). The values were mostly close to the detection limit.

No endosulfan residues were found in human breast milk and urine and no serious clinical effects were detected in the local population after monitoring.

Domestic animals were also not affected by endosulfan application.

As anticipated from the results of previous studies, the Scientific Environmental Monitoring Group (SEMG), which carried out the spraying operations in Zimbabwe in 1986, found that there were no (or only very minor) deleterious effects on terrestrial non-target organisms, including man and domestic animals. Such effects that were recorded were of little significance in relation to the other factors causing habitat disturbance, such as subsequent land utilization, bush fires, deforestation, other pesticides and climatic variations.

The effects on stagnant water bodies such as dams and pools were very small; they usually overlapped the natural fluctuations of the fish population. As expected, fish deaths occurred in running water systems; however, at no study site was there any absence of fish. At the conclusion of spraying, different species of live fish were recovered from all the water bodies examined. However, the absence of long term effects on fish populations cannot be assumed from these observations, and further study is necessary.

On the basis of these results, the SEMG considers that in areas comparable to those sprayed in the 1986 control operation in NE Zimbabwe endosulfan applied in the same manner as the sequential aerosol spraying technique is *environmentally acceptable*. In future control operations, extreme caution should be exercised with regard to running water systems, and particular attention should continue to be paid to the impact on fish populations.

Compared with the helicopter residual spraying in Cameroon, the ecological effects resulting from the non-residual sequential aerosol spraying technique were negligible.

2.4. Impact on and residues in ecosystems and humans

Any valuable ecological monitoring will only be feasible if ecotoxicological studies are linked with the whereabouts of the noxious agent in the food web [1, 17,

TABLE III. RESIDUE LEVELS OF DIELDRIN IN ANIMAL LIVERS IN SPRAYED AREAS (mg dieldrin/kg fresh weight)

		Months after treatment					
	No.	1	8	11	23	35	72
Aves (insectivores)							
Ispidina picta	27	0.6	—	0.58	0.62	0.43	0.017
Halcyon malimbica	30	4.3	5.0	6.52	2.24	0.6	—
Aves (omnivores)							
Turdus pelios	29	0.44	2.08	3.0	0.075	0.045	0.001
Macrochiroptera							
Micropterus pusillus	24	136.0	—	0.14	0.075	0.065	0.0
Microchiroptera							
Hipposiderus commersoni	10	3.9	—	—	0.11	0.185	—
Soricidae	6	11.3	2.3	—	—	—	0.04
Rodentia	27	0.35	0.35	0.24	0.065	0.085	0.006

18]. Apart from the size of the habitat, the migratory character of some animal and bird species will also play an essential role in interpreting the results obtained from residue analyses [1, 18].

In view of the fact that large parts of Adamaoua had already been treated with dieldrin before spraying of the gallery forest began, it was necessary to check the dieldrin content of various species. Directly after spraying, there was a strong variation in the diverse trophic levels and individual groups investigated. This can be explained by the varying exposure of individual specimens, their different behaviour and their different habitats.

The highest contamination was detected in fructivorous bats directly after spraying, whereas the residue values for insectivorous alipeds were considerably higher only after 2 or 3 years. The residue levels of bird species, which were already low in 1981, remained constant in 1982.

The residue values in insectivorous species such as Ispidina picta and Halcyon malimbica were still very high in 1981 but there was a marked drop in the level between 1981 and 1982. This can be explained by the fact that these birds feed mainly on insects and prefer a savanna habitat.

The standard residue values found in four bird species, namely Ispidina picta, Halcyon malimbica, Indicator indicator and Cossypha albicapilla, were the reason why a mean for all trophically similar organisms was calculated.

Table III and Fig. 4 show the characteristic residue trends for dieldrin with regard to various genera; they are valid for the period 1979 to 1983.

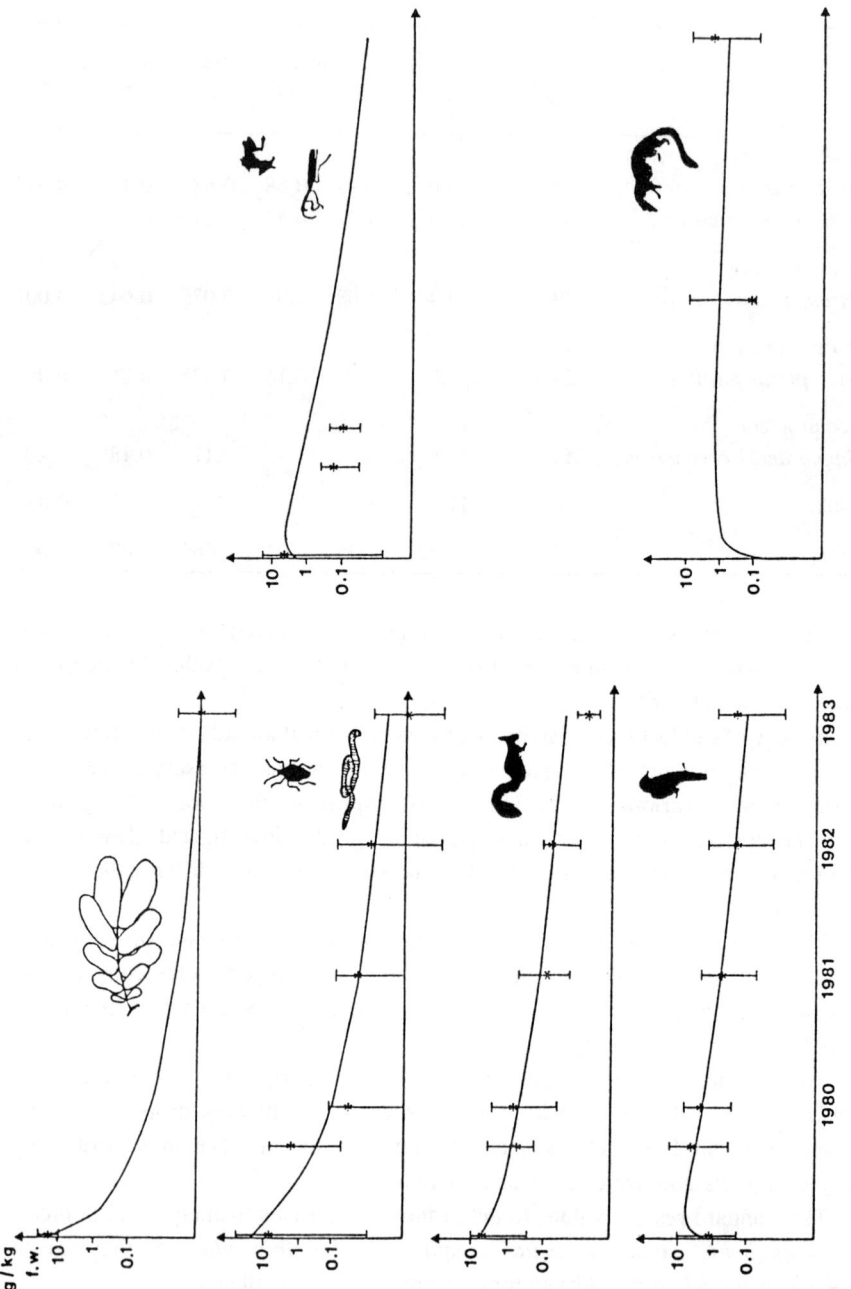

FIG. 4. Dieldren residues in different components of the food web from the treated area in Cameroon [1].

TABLE IV. ENDOSULFAN RESIDUE LEVELS IN DIFFERENT COMPONENTS OF THE FOOD WEB FROM THE TREATED AREA (1986). THE DDT LEVELS OF THE SAMPLES ARE ADDED FOR COMPARISON PURPOSES (mg/kg)

	Tissue	Date	Location	X days after cycle Y	Endo-sulfan	Total DDT
Catfish (maximum level)	bo	11.8	NyR	1/II	5.10	8.05
Honey bees (n = 8)	bo	11.9	Ku	0/IV	0.100	0.013
Lycophidion capense	bo	14.9	Ku	3/IV	0.085	0.175
Anura	bo	28.8	Ku	1/III	0.045	0.024
Rodent No. 158	li	27.7	Ku	7/I	0.010	0.054
Amaurornis flavirostris	li	2.8	Ku	12/I	0.004	0.066
Lybius torquatus	br	26.8	Nya	18/II	0.004	0.006
Mabuya species	li	19.7	Ku	0/I	0.003	0.010
Hieraetus fasciatus	fe	12.9	Mu	1/IV	0.003	1.84
	li				nd	5.25
	fat				nd	723
Rodent No. 154	li	26.7	Ku	6/I	0.003	0.014
Cow milk	mi	9.8	Ku	1/II	0.003	0.009
Goat milk	mi	10.8	Ku	2/II	0.002	0.011
Elephant shrews	li	27.7	Ku	7/I	0.001	0.015
Amaurornis flavirostris	li	1.8	Ku	11/I	0.001	0.061

bo = whole body; li = liver; br = brain; fe = feather; fat = body fat; mi = milk (fresh weight); NyR = Nyadiri River; Ku = Kudzwe Dam; Nya = Nyamatawa; Mu = Mudzi River; nd = not detectable.

In 1986 approximately 1200 wildlife samples were collected in different parts of Zimbabwe (the Makuti, Rukomeche, Kariba, Mt. Darwin areas). The basic data occasionally showed very high DDT residue levels but very low or no residues in other halogenated hydrocarbons. No endosulfan residues were detected in the pre-spray samples.

All the fish samples were collected after a single treatment in the Nyadiri River and the Kudzwe Dam. The endosulfan detection limit was 0.001 mg/kg. All the specimens analysed to date have shown insecticide residues ranging from 0.055 mg/kg in the liver of living species up to 5.1 mg/kg in the liver of dead specimens, the latter obviously being the cause of death (Table IV).

TABLE V. RESIDUE LEVELS OF TOTAL DDT AND ENDOSULFAN IN THE LIVERS AND EGGS OF CHICKENS FROM THE TREATED AREA (KUDZWE DAM) (mg/kg fresh weight)

	Total DDT				Endosulfan
	Minimum	Maximum	Mean	Median	
Liver (n = 24)	0.925	25.5	6.95	4.7	nd
Egg yolk (n = 12)	3.80	53.0	18.9	12.0	nd

The condition indices of omnivorous poultry and their gonad development did not show any significant difference between sprayed sites and the control area. Those of a small herd of cattle (muscle profile and adipose cover) did not reveal any difference between investigation periods or between sprayed and unsprayed areas.

Poultry eggs are a regular part of the food of the local population and were therefore analysed for residues of halogenated hydrocarbons (Table V). In none of the egg yolks analysed has endosulfan been found; it was also not present in the chicken livers. In contrast, the DDT levels were very high in all samples, especially in the egg yolks.

Residues of endosulfan were found in cow milk (0.003 mg/L fresh weight, 0.160 mg/kg lipid weight) as well as in goat milk (0.002 mg/L fresh weight, 0.030 mg/kg lipid weight). These levels are above the acceptable daily intake, which is 0.008 mg/kg body weight for endosulfan.

Over the whole spraying period residue samples of breast milk and urine were taken from people living in block 1. Forty-six samples of human breast milk were analysed but no endosulfan residues were detected (detection limit: 0.02 mg/L, based on lipid weight) (Table VI). Forty-four urine samples were analysed but no endosulfan residues were detected (detection limit: 0.0001 mg/L). The levels of the other halogenated hydrocarbons were very low.

3. ELIGIBILITY OF INSECTICIDES FOR USE IN AERIAL SPRAYING OPERATIONS

Detailed investigations over many years have shown that endosulfan can be effective against *Glossina* species and is environmentally accceptable when sequentially applied as a low dosage aerosol. However, we should recognize the potential

TABLE VI. RESIDUE LEVELS IN HUMAN BREAST MILK (mg/kg lipid weight)

	Minimum	Maximum	Mean	Median
Alpha-HCH	nd	0.90	0.11	0.02
Beta-HCH	nd	4.20	0.77	0.60
Lindane	nd	0.10	0.02	0.02
Total DDT	3.45	81.5	35.1	32.0
Endosulfan	nd	nd	nd	nd

value of alternative insecticides, e.g. for aerial spraying. For new insecticides the following criteria are acceptable:

(1) The active ingredient must not be banned from use in any country worldwide
(2) There must be no adverse effects on human health regarding the formulation and the application rates proposed
(3) There must be no adverse effects on the biospheric processes
(4) The mammalian, bird and fish toxicity, as demonstrated by standard tests, must normally not exceed that of endosulfan at the expected dose rate and formulation
(5) The insecticide must have demonstrated the absence of effects likely to be detrimental to the environment; field testing must be conducted under the aegis of a professional team
(6) The insecticide must be formulated for application by a rotary atomizer operating at 10 to 12 000 rev/min to give a droplet spectrum in the volume median diameter (VMD) range of 20 to 30 μm (measured on 0.64 cm MgO coated slides) at the target site when applied from heights between 25 and 50 m
(7) The insecticide must be as effective as endosulfan for the control of *Glossina* when applied as an aerosol in the VMD range of 20 to 30 μm
(8) The formulated insecticide must be economically competitive with existing alternatives, except when its use gives rise to a significant reduction in environmental risk or operator hazard, or otherwise represents an advance in the capability of aerial spraying.

The protocol for insecticide approval must include:

Phase I: Toxicity data for tsetse flies

Laboratory determination of toxicity should be done by precise application of the aerosol in the required droplet range to give a known dosage to adult *G. morsi-*

tans and *G. pallidipes* within the temperature range of 4 to 35°C. Data must be presented on insecticide toxicity to flies which cover the full range of age, nutritional and reproductive status normally encountered in a wild population.

Phase II: Small scale field trials

Aerial application by the fixed wing aerial spraying technique, or an alternative method, to an agreed block of at least 400 km^2 to test the suitability of the formulation, to assess the efficacy of the insecticide against the target insect under field conditions and to monitor the impact of insecticide application on non-target organisms should be carried out according to procedures agreed to beforehand.

Phase III. Large scale field trials

Our experience in tropical South America, Africa and Europe has proved that the application of chemicals does not represent a serious environmental problem, provided the chemical inputs on the ecosystems have already been evaluated, and the whole procedure and organization are carried out in a professional manner [19]. Concerning the elimination of pests, it has been observed that various forms of land utilization in the tropics have a much more adverse effect than those effects induced by pesticide usage.

REFERENCES

[1] MÜLLER, P., Ecological side effects of dieldrin, endosulfan, deltamethrin and cypermethrin application against tsetse flies in Adamaoua, Cameroon, World Bank, Washington, DC (1987) 1–194.
[2] WINTERINGHAM, F.P.W. (Ed.), Environment and Chemicals in Agriculture, Proc. CEC Symp., Elsevier, Amsterdam (1985) 407.
[3] FOOD AND AGRICULTURE ORGANIZATION OF THE UNITED NATIONS/WORLD HEALTH ORGANIZATION, Pesticide Residues in Food — 1985, Plant Production and Protection Paper No. 68, FAO, Rome (1986) 78.
[4] HILL, J.R., et al. (Eds), Pesticide Microbiology, Academic Press, New York (1978) 844.
[5] BÜCHEL, K.H. (Ed.), Chemistry of Pesticides, Wiley, New York (1983) 518.
[6] MÜLLER, P., Arealsysteme and Biogeographie, Ulmer, Stuttgart (1981) 1704.
[7] MÜLLER, P., Zur Rückstandssituation bei freilebenden Tieren in der Bundesrepublik Deutschland, Mitt. Biogeographie Saarbrücken **15** (1985) 54.
[8] CADOGAN, B., et al., Spray deposits and drop size spectra from a high wing monoplane fitted with rotary atomizers, Trans. Am. Soc. Agric. Eng. **29** (1986) 402–406.

[9] BOUSE, L., CARLTON, J., Factors affecting size distribution of vegetable oil spray droplets, Trans. Am. Soc. Agric. Eng. **28** (1985) 1068-1073.

[10] RICHARDS, R.P., et al., Pesticides in rainwater in the northeastern United States, Nature (London) **327** 6118 (1987) 129-131.

[11] WORLD METEOROLOGICAL ORGANIZATION, The Global Climate System, WMO, Geneva (1987) 87.

[12] GEORGHIOU, G.P., SAITO, T. (Eds), Pest Resistance to Pesticides, Plenum Press, New York (1983) 809.

[13] WILSON, R., et al., Risk assessment and comparisons, Science **236** 4799 (1987) 267-300.

[14] MÜLLER, P., Wissenschaftliche Grundlagen and normative Wirkungen von Grenzwerten, Acad. Humb. Nova (in press).

[15] MÜLLER, P., NAGEL, P., Environmental Monitoring of Tsetse Control Operations in Zimbabwe, European Economic Community, Brussels (in press).

[16] JORDAN, A.M., Trypanosomiasis Control and African Rural Development, Longmans, Harlow, Essex (1986) 357.

[17] SHEENAN, P., et al. (Eds), Appraisal of Tests to Predict the Environmental Behaviour of Chemicals, SCOPE Monogr. No. 25, Wiley, New York (1985) 380.

[18] FORD, M.G., et al. (Eds), Neuropharmacology and Pesticide Action, Ellis Horwood, Chichester (1986) 512.

[19] WARE, G.W. (Ed.), Reviews of Environmental Contamination and Toxicology, No. 100, Springer-Verlag, Berlin (West) (1987) 156.

IAEA-SM-297/42

Invited Paper

CHEMICAL AND RADIOACTIVE RESIDUES IN SOIL
A global perspective

F.P.W. WINTERINGHAM
Knoll Hill, Sneyd Park,
Bristol, United Kingdom

Abstract

CHEMICAL AND RADIOACTIVE RESIDUES IN SOIL: A GLOBAL PERSPECTIVE.
    Attention is drawn to the current global trends in population, agrochemical usage, chemical and radioactive emissions, and the appearance of an increasing range of anthropogenic chemical and radioactive residues in soils worldwide. The consequent exposure of soil organisms, plants and dependent populations, simultaneously or consecutively, to such multiple residues emphasizes the need for a more comparative and integrated approach to their study and appraisal. This would provide a safer base for priority allocation and decision making for the protection of environmental quality, biological resources and public health. It would also assist public understanding of relative hazards and the importance of seeing a particular environmental problem in context.

1. INTRODUCTION

    This symposium is being jointly held with the 10th Biennial Meeting of the International Academy of Environmental Safety and the Society of Ecotoxicology and Environmental Safety (SECOTOX). International co-operation of this kind has never been more important in today's world of escalating technology, chemical and radioactive usage, and emissions.
    In an attempt to view these problems globally and impartially, some relevant trends and implications are indicated without detailed data. These data and their supporting evidence are identified in the references cited in the printed text of this and earlier reviews (e.g. Refs [1–3]). This obviates the need for too many tables of data already published.
    Global population is increasing, as well as chemical and energy use per head. They are also extending geographically with so-called development. Meeting global demands for food and other plant products increasingly depends upon the intensification of agriculture, forestry and fisheries. These, in turn, involve rising levels of energy use per unit of energy harvested, mechanization, artificial irrigation and

agrochemical usage. We are all surely aware that the occasional and relatively local surpluses of harvested biomass are no indication that we can afford to 'turn the clock back' in relation to these intensified practices except, of course, by accelerating net deforestation and the cultivation of more of the effectively dwindling land resources per head. Moreover, these current trends are associated with emerging and sometimes serious constraints: soil erosion, spreading deserts, rising marine and atmospheric pollution, and the extending and sometimes accumulating residues in cultivated soils worldwide.

Despite this gloomy scenario there is, fortunately, an increasing number of responsible scientists drawing attention to the already urgent need for planetary management and the necessary international scientific communication and co-operation implied (e.g. Refs [4, 5]). Hopefully, they "will exert the necessarily sufficient and timely influence at the international level" [6].

2. RESIDUE PROBLEMS IN SOIL

2.1. Importance of the comparative and integrated approach

There is no doubt that for food, timber, many fibres and, indeed, for fuel for large populations soil resources and their quality are vital to human survival in the current state of knowledge.

Man and his environment are exposed to many contaminants, either consecutively or simultaneously, during a lifetime. Therefore, ecotoxicologists have a particular responsibility for presenting their information and data in a comparative and integrated context [7]. There is, otherwise, a serious danger that popular media presentation may lead to the wrong decisions or to the allocation of the wrong priorities in relation to both environmental safety and public health. Media handling of the Chernobyl nuclear accident provides a conspicuous example [8].

The term residue is used here in the sense of any chemical, whether radioactive or not, which appears in the soil as a result of human activity and where or when it no longer serves a useful purpose. It can, therefore, be a natural substance such as nitrate moving below the root zones because of fertilizer application and leaching. Biological residues such as those resulting from the use of genetically engineered strains of bacteria or enzyme preparations for pest control, soil decontamination, etc. [9] are not discussed here.

Residues in soil may result from on-site agricultural practices such as agrochemical usage, irrigation, and as a result of drift or fallout from distant practices. Both chemical and radioactive residues appear in soil as a result of dry or wet deposition from remote point, line, or area sources, e.g. radioactive fallout from a nuclear accident or weapons testing (point source); acid rain from an industrial area

(area source); lead from an adjacent motorway, or salt from a flooding tidal river (line source).

Several important questions arise:

(1) What are the origins, trends and international implications of a soil residue?
(2) What are the physical and chemical fate of a residue in real time?
(3) What is the significance of the residue in terms of the undesirable effects on soil quality, fauna, flora and microflora?
(4) To what extent does it become transferred into an edible crop or into groundwater or derived surface water?
(5) Finally, and of greatest importance, what is the human health significance of a residue moving into feed, food or drink?

There is an immense amount of literature available on these questions and only a brief summary is attempted here as an aid to 'not losing sight of the forest for trees'. Since the first United Nations conference on the human environment merely 15 years ago, the universe of publications in environmental science seems already to have expanded beyond even the scholar's telescope.

2.2. Trends, transport and implications

The rising and extending use of agrochemicals is well documented, and seems certain to continue into the next century [3, 10]. Despite controls in some industrialized countries, there are also extending and rising emissions from industry, transport, heating, urban and industrial wastes, etc., especially in developing countries of tropical and subtropical latitudes [11].

An interesting and important aspect of environmental pollution is the critical role of weather in atmospheric transport. This carries residues across international boundaries and results in deposition on a global scale, as demonstrated by the earlier testing of nuclear weapons in the atmosphere and by the recent Chernobyl accident [12].

An example of long range atmospheric particulate transport was the recent deposition of sand from the Sahara Desert over much of the United Kingdom in August 1987 which prompted a day of nationwide car washing [13]! The presence of insecticide and herbicide residues in rainfall has demonstrated transport in solution [14]; transport either in solution or suspension in fog aerosols has also been observed [15].

Atmospheric transport may not only result in new soil residues but, in the form of acid rain, may also bring about a significant shift in the pH of soil and aquatic ecosystems. This can, in turn, affect the behaviour of existing nutrients and residues [16]. However, and incidentally illustrating the importance of the comparative approach, soil nitrification in fertilized agricultural soils may be a far more important factor in the soil pH [17].

These considerations illustrate the international implications of soil contamination and the corresponding communication and co-operation needed for their effective control, as recently stressed by C. Davis of the Commission of the European Communities [18].

2.3. Physical and chemical fate in soil

The time–concentration curve for a specific residual element or compound in, for example, the critical root zone of the soil will clearly depend upon: (1) the nature, magnitude and frequency of the input, and (2) the effective disappearance rate as a result of the combined abiotic and biotic removal mechanisms, including that of radioactive decay in the case of radioactively significant residues. These will also include physical loss from the surface by volatilization, downward leaching, biological uptake by the plant, microbiological degradation, etc. [19].

Inputs may be characterized as continuous or discontinuous: for example, input over pasture exposed to chemical fallout (from distant prevalently upwind industrial areas) or radioactive fallout (such as that from the original atmospheric nuclear weapons testing), and when disappearance rates are relatively slow, would be continuous; the annual application of a fertilizer or pesticide to a cultivated soil would be discontinuous; a single but very significant input of relatively short duration (such as the initial fallout episodes following the Chernobyl accident) would be a 'spike' input.

As discussed in Ref. [19], the significant implication is that, for comparable time weighted mean input rates, acute ecotoxicological effects would be expected to be higher for the discontinuous input, other things being equal.

Changes in the chemistry of a soil residue have also been the subject of extensive studies. Because of the well established and often critical role of isotopic tracer techniques, several relevant programmes have been co-ordinated by the Joint FAO/IAEA Division of Isotope and Radiation Applications of Atomic Energy for Food and Agricultural Development in Vienna (e.g. Refs [20–23]).

Chemical modification will be due to abiotic factors such as photochemical, oxidative, reductive and hydrolytic ones, and to biotic factors such as plant uptake and metabolism, soil microbiological absorption and metabolism, etc. The organic matter content, fertility and microbiological population are especially important in the potential for biotic degradation. For this reason, use of ^{14}C labelled substrates as an indicator of the biotic degradation potential of aerobic soils has been suggested [20]. This approach is well established as a primary productivity indicator for aquatic ecosystems (e.g. Refs [22, 23]).

2.4. Behaviour and ecotoxicological significance

Many models have been described for rationalizing, simulating and predicting the movement of a plant nutrient, chemical or radioactive residue in the soil–plant system, and in derived foodwebs (e.g. Refs [19, 24–26]). It is sufficient to reiterate here that post-Chernobyl experience has confirmed that models are no substitute for on-site and real time monitoring where there is an actual or suspected threat to public health [8, 12].

This observation is not intended to imply that models in the form of mathematical and computer aided simulation, or in the form of a miniature laboratory ecosystem, lysimeter, etc., are not extremely useful and powerful tools, especially for providing an integrated quantitative picture of the expected behaviour or comparative behaviour and significance of a residue. It is simply to recognize that the deposition, movement and behaviour of a residue may depend, critically, upon local weather at the time of addition or deposition and, afterwards, upon the status of the soil and floral cover, and upon the soil biotic activity. Local weather remains notoriously immune to accurate prediction and the complexities of the soil–plant system, associated biotic populations and their constant changes make accurate prediction difficult.

There is also a great deal of literature available on the biochemistry and toxicology of environmental chemical and radioactive residues (e.g. Refs [8, 12, 26–30]), not to mention a continuing series of publications on limits, etc. for human exposure and intake (e.g. Refs [31, 32]).

In relation to behaviour and toxicology there are important differences between a chemical and radioactive fallout residue in the soil. Because of its effectively carrier free or isotopically undiluted condition, the chemical concentration of a fallout radionuclide may be extremely low. This is illustrated in Table I.

The downward leaching of a chemical residue, especially that of a soluble anion such as nitrate or chloride, can often be explained on the basis of partition chromatographic elution when the residue behaves as a solute partitioned between an effectively stationary solvent phase and the mobile phase of infiltrating water [25]. The extremely low concentration of a fallout radionuclide, however, suggests very high adsorption on soil particulate and structural surfaces on the basis of the classical adsorption theory, at least initially, or until effectively diluted or displaced by the natural isotope (if present) or by suitably chemically related ionic species [12], or, additionally, until possibly complexed and mobilized by microbiological action or reaction with the organic components of the soil [33]. This probably accounts for some of the difficulties observed after Chernobyl in attempts to decontaminate freshly harvested fruit, vegetables, etc. by simple washing; also for the often very slow migration of fallout radionuclides in the soil profile. Those familiar with the task of decontaminating, even glassware, after containing carrier free radionuclides will be painfully aware of these problems.

TABLE I. CHEMICAL CONCENTRATIONS OF FALLOUT RADIONUCLIDES EQUIVALENT TO A DEPOSIT OF 10 000 Bq·m^{-2} OF SOIL TO A DEPTH OF 10 cm AND AN ASSUMED BULK DENSITY OF 1.5 g·cm^{-3}

Radionuclide	Concentration (μg·g^{-1} (ppm))
Strontium-90	1.3×10^{-8}
Ruthenium-106	5.3×10^{-10}
Iodine-131	1.3×10^{-11}
Caesium-137	1.7×10^{-8}
Plutonium-239	3×10^{-5}

The biological effects of a radionuclide will be entirely due to the nature and level of the emitted radiation (or to the chemical transmutation on decay under rare circumstances). They would not be a function of chemical concentration per se. Plutonium could be an exception because of the relatively large weights accumulating in a reactor core and its high chemical and radiotoxicity. Chemical modification of a radionuclide will, of course, affect its behaviour and effective half-life in the organism but not its radiotoxicity. A characteristic of a radioactive residue on or near the soil surface is that high levels could represent an external radiation hazard (e.g. for agricultural personnel working on the land). All these aspects have been discussed and illustrated elsewhere [8, 12].

The ecotoxicology of a chemical residue will be strictly a function of its biochemistry, so that biotic absorption and metabolism may decisively modify the toxicity of the molecule, while radiotoxicity at a particular intracellular site will not be so affected.

Another aspect is that, in relation to human health, early somatic or late stochastic effects in a very small fraction of the exposed population may be of societal significance. In relation to soil organisms and wildlife, only obvious changes in populations are likely to be of concern. Moreover, the well established phenomena of pest resistance to biocides indicates the relative adaptability of wildlife populations in a toxic environment. However, an important societal consequence of pest resistance itself is the persisting need to develop and evaluate alternative chemicals, formulations and pest management techniques [3].

Regarding the ecotoxicology of radioactive residues, because of extreme public sensitivity to the possibility of their appearance in food or drink and also the relatively lower radiation response of microorganisms and wildlife [12], the following concensus has emerged: it would be "the impact on man" that would be "important rather than the effects on other components of the biosphere" [26].

Finally, when appraising the environmental or agricultural significance of a particular biotic–soil residue interaction it is very important to view it in context. For example, while a pesticide residue may, indeed, disturb or poison some microbiological population in the soil, this may be of minor significance compared with the loss of fertility due to some other ongoing agricultural practice such as clearance with burning and erosion, unsuitable irrigation water, etc. [6].

3. PROBLEMS AND NEEDS

3.1. Problems

Against this background some relatively neglected problems and international needs can be recognized.

Residues in soil as a result of agricultural or forestry practice can usually be anticipated and controlled at the farm or national level. Unintended residues as a result of fallout from distant routine or accidental emissions, on the other hand, may neither be detected nor anticipated, and effective control may involve national legislation and international agreement. Pesticide and irrigation water residues are examples of the former, while acid deposition, toxic metals and radioactive fallout are examples of the latter.

However, as already mentioned, long range atmospheric transport of remotely applied pesticides and contaminated soil particulates is now recognized. Likewise, fertilizer derived nitrates reaching rivers may cross international boundaries. Moreover, soil residues generated by local practices may become transferred into feed or food products which enter international trade. This may, additionally, involve the use of an agricultural chemical not approved by the importing country. In short, soil residues are increasingly becoming international problems in terms of movement, acceptability and control [3]. This has been dramatically demonstrated in the case of radioactive residues by the Chernobyl experience [8, 12].

A general problem of environmental safety is the continuing harnessing of science and its derived technologies to commercial, national and other unilateral interests, including the scientific disciplines themselves. The problem of expanding information and data beyond intellectual absorption capacity has been mentioned. This itself hinders the international scientific communication and co-operation now obviously essential to improve environmental protection.

Recognition of this danger some years ago prompted the first and only meeting in Vienna which brought together scientists concerned both with pesticide and radioactive residues [34].

Within the European Economic Community and elsewhere [2, 3] the problem of fertilized soil nitrate reaching drinking water supplies is now recognized as a serious and growing one. This problem is clearly linked to the obsession with agricultural yield maximization rather than optimization, which would also take into account resource and environmental quality protection. This, in turn, raises sensitive but environmentally critical questions about the wisdom of uncontrolled market force competition and rivalries. What is clear is that any effective solution will again depend upon international goodwill and co-operation. Likewise, it is clear that any effective shift towards more *sustainable* agricultural, forestry and fisheries practices must involve greatly improved international communication and co-operation.

A related problem is the proliferation of terminology, units, limits, etc. which becomes confusing, even for scientists. This, again, is due to the lack of sufficiently wide international scientific consultation, communication and co-operation. For example, the Food and Agriculture Organization of the United Nations/World Health Organization and related Codex Alimentarius Commission programmes on pesticide residues carried out over the last decade have employed and defined such terms as unintentional residue (later dropped), practical residue limit (superseded by extraneous residue limit), acceptable daily intake (ADI), conditional ADI, temporary ADI, tolerance (superseded by maximum residue limit), guideline level, maximum limit for pesticide residue, etc. [35–47].

The Chernobyl accident has exposed the continuing use of many terms and different units — often relating to the same parameters. Some examples are: intervention level, derived intervention level, reference level, protective action guide (PAG), preventative PAG, derived response level (DRL), preventative and emergency DRLs, maximum tolerance, interim international radionuclide action level for food (IRALF), not to mention becquerel, gray and sievert in addition to the classical units of curie, rad and rem. The latter, of course, are well established after almost 50 years of publications on environmental radioactivity and radiological protection [8].

These remarks are not meant to assign blame. They have, however, been made to illustrate a growing problem and to suggest that the dropping of obsolete or superfluous terminology and units can be useful, but to invent new ones after decades of published use is not necessarily progressive or scientifically justified.

3.2. Needs

The importance of the comparative approach has been stressed in relation to public health and environmental quality protection. Under the almost invariable conditions of simultaneous or consecutive exposure to two or more residues it will be the overall net impact that matters.

Two or more residues (whether radioactive or not) can act jointly, so that the net impact may be the result of additive, less than additive (antagonistic), or more than additive (synergistic) mechanisms. This underlines the need for the integrative approach [38].

The comparative and integrative approach implies the need for sufficient comparable information about each potentially significant residue in terms of behaviour and ecotoxicology.

The preparation and publication of the Joint FAO/IAEA Summaries was one attempt to meet this need more than a decade ago [39]. A questionnaire at the time indicated a unanimous welcome by research workers of developing countries who often lacked access to the sophisticated libraries and information services of the industrialized countries. These research workers nevertheless faced growing environmental problems of their own. Each Summary was always limited to two pages and dealt with a single residue such as 2,4-D [40], or tritium [41], or residues by group such as all pesticides [42] or all radionuclides [43].

The unilateral, as opposed to the comparative and integrated, presentation of data can lead to unsound judgement and unwarranted public fear. The stochastic risks of late cancer as a result of the Chernobyl fallout illustrate the problem.

Stochastic radiation risks of late cancer are usually based on a projected period of 50 years or more, after commencement of the significant radiation exposure. In the case of absorbed radionuclides, the estimates take into account the committed integrated radiation dose expected on the basis of effective half-life, expected distribution within the body, etc.

Table II illustrates the trends in the cause of death in the UK and relevant longevity data [44–46]. The data are probably typical of Europe as a whole. It has been authoritatively reported [47] that additional cancer deaths within the EC as a result of Chernobyl will be of the order of 1000 over the next 50 years. On the basis of Table II and the present EC population, the Chernobyl impact will cause less than one additional case per 50 000 fatal cancers. In relation to public health priorities it is, perhaps, pertinent to note that fatal road accidents in the same countries over the same 50 year period can be expected to be of the order of at least 2 million [48].

It is also recognized that there is a major environmental factor in the occurrence of all cancers, of which a significant proportion can be assigned to environmental or dietary chemicals [46, 49, 50]. The important subject of comparative stochastic risk assessment has recently been discussed [49].

Reference has been made to the emphasis that has been placed on agricultural yield maximization at the expense of resource depletion and environmental quality rather than optimization, taking all these factors into account. This emphasis has continued to dominate recent and otherwise useful and constructive scientific symposia (e.g. Refs [51, 52]).

Regarding soil residue problems, many R&D needs can be identified (for model development, see Section 4). Time and space dictate mention of only a few here.

In studying the effects of residues or agrochemical usage the well established phenomena of population acquired tolerance and even immunity to toxic chemicals

TABLE II. LONGEVITY AND CAUSE OF DEATH, UNITED KINGDOM [44–46]

	Average life expectancy (years)		
In the year:	1900	1960	2000
All males	49	68	>71
All females	52	74	>77
	Cause of death (per cent all deaths)		
In the year:	1900	1960	1978
Infectious disease	35	12	10
Heart disease	14	38	34
Cancer	5	19	22
Other	46	31	34
Total	100	100	100

indicate the need to give more attention to the effects on populations over a sufficient period of time or number of generations. Thus, the use of soil nitrification inhibitors may become seriously constrained by the emergence of inhibitor resistant microorganisms [53]. Conversely, for similar reasons, laboratory or model ecosystem toxicity studies may not be a reliable guide to effective toxicity over long periods of time under field conditions [6].

Such accelerated genetic selection, on the other hand, may confer greater tolerance of useful plant species, as observed in the resistance of higher plants to toxic metal residues [54]. This, in turn, raises the interesting question as to whether the transfer of residues from the soil to the edible parts of the plant is also higher and suggests a useful extension of studies made of the movement of heavy metals into the food chain [55].

In this same connnection it has also been established that soil microorganisms may not only degrade an otherwise toxic pesticide residue but thrive on it as a new substrate, especially under the anaerobic conditions of a flooded rice paddy (see Sethunathan, Paper 5.2, and Magallona, Paper 5.3, in Ref. [11]). These and earlier observations suggest potential for the cultivation of resistant microbial populations for the decontamination of soils or, indeed, for the use of biotechnically based enzyme preparations for the same purpose [9].

The Chernobyl experience has indicated some anomalies in the transfer of fallout radionuclides from soil to plants, and in the effectiveness of decontaminating harvested vegetable crops by simple washing. This suggests the need to study further the probably critical effects of the time factor, and of the very low chemical concentrations involved [8, 12], in addition to the normal parameters studied, e.g. under conditions otherwise comparable with those for studying the behaviour of herbicides [56].

4. CONCLUSIONS

Rising and/or extending chemical and radioactive residues in soils worldwide indicate the growing importance of comparative and integrated studies. This can be the only logical basis for the allocation of priorities and for optimal decision taking in the interests of public health and environmental quality and resource protection. While a single residue might, indeed, adversely affect the natural population of an agro-ecosystem, it may be of minor importance compared with other changes taking place in the agro-ecosystem [57].

The presentation of appropriate collections of ecotoxicological data in a more comparative and integrated format would also help understanding by the non-specialist and facilitate the appraisal of a particular problem in context. The term appropriate here refers to the ecosystem under consideration, likely exposure pattern, agricultural practices, location (proximity to possible emissions, etc.) and climate.

Ecotoxicological profile analyses [58] are a major step in the right direction and recent publications indicate recognition of this direction at the scientific level [19, 29, 57, 59, 60]. In developing the comparative approach it will be useful not to lose sight of the existing mass of information and data on individual elements, compounds and radionuclides contained in earlier reviews, e.g. one prepared as an FAO/UNEP report [61], which might be available on request to the Environment and Energy Programmes Co-ordination Centre, FAO, Rome. A second useful review in the context of climate is in preparation as a SCOPE monograph [11]. Two post-Chernobyl reports also review and summarize key literature on the behaviour and significance of radioactive fallout in agro-ecosystems as studied during the atomic energy era [8, 12].

Recent experience [8, 62] has confirmed the need for caution when using laboratory experiment based models for predicting the behaviour of soil residues, whether they are radioactive or not. On the other hand, integrated models have already become an essential tool for handling the escalating data on residues, their behaviour and ecotoxicology, and for indicating relative hazard potential [63]. These developments will also facilitate the complex tasks of realistic legislation and controls at state and national levels, given the necessary additional information and data needed. In particular, there remains a need for "integrated field scale experiments in which all the important parameters governing atmospheric and soil fate are simultaneously monitored" [62].

Finally, scientists themselves carry an increasing societal responsibility for improved environmental safety on the now obviously needed international scale. The nature and complexity of the problems are clearly beyond the scope of conventional diplomatic machinery and unilateral national actions (see Ref. [9]).

REFERENCES

[1] WINTERINGHAM, F.P.W., Biomass cultivation and harvest: Global trends and prospects, Outlook Agric. **12** 1 (1983) 21–27.

[2] WINTERINGHAM, F.P.W., Soil and Fertilizer Nitrogen, Technical Reports Series No. 244, IAEA, Vienna (1984).

[3] WINTERINGHAM, F.P.W. (Ed.), Environment and Chemicals in Agriculture, Proc. CEC Symp. 1985, Elsevier, Amsterdam and New York (1985).

[4] MYERS, N., The Gaia Atlas of Planet Management, Pan Books, London (1985).

[5] MALONE, T.F., ROEDERER, J.G., Global Change, Proc. ICSU Symp. 1985, Cambridge University Press, Cambridge (1985).

[6] WINTERINGHAM, F.P.W., Environmental toxicology in relation to biomass cultivation and harvest, Rev. Environ. Toxicol. **1** (1984) 1–4.

[7] WINTERINGHAM, F.P.W., Foreign chemicals and radioactive substances in food and environment: A comparative and integrated approach to the problems, Kem. Teollisuus **29** 9 (1972) 561–574.

[8] WINTERINGHAM, F.P.W., Behaviour and Significance of Radioactive Substances Released into Agricultural, Forestry and Fisheries Ecosystems, Part I of a Background Review with Particular Reference to Accidental Releases, Chernobyl and the Future, IAEA, Vienna (in preparation).

[9] BATRA, L.R., KLASSEN, W. (Eds), Public Perceptions of Biotechnology, Agricultural Research Institute, Bethesda, MD (1987).

[10] FOOD AND AGRICULTURE ORGANIZATION OF THE UNITED NATIONS, Agriculture: Towards 2000, Conf. Doc. C.79/24, FAO, Rome (1979).

[11] BOURDEAU, Ph., et al. (Eds), Ecotoxicology and Climate, SCOPE Monogr., Wiley, New York (in press).

[12] WINTERINGHAM, F.P.W., Soil and Crop Contamination by Radioactive Fallout, Part II of a Background Review with Particular Reference to Accidental Releases, Chernobyl and the Future, IAEA, Vienna (in preparation).

[13] Press report, The Times, London (17 Aug. 1987) 6.

[14] RICHARDS, R.P., et al., Pesticides in rainwater in the northeastern United States, Nature (London) **327** 6118 (1987) 129–131.

[15] GLOTFELTY, D.E., et al., Pesticides in fog, Nature (London) **325** (1987) 602–605.

[16] WORLD METEOROLOGICAL ORGANIZATION, The Global Climate System, WMO, Geneva (1987).

[17] ENVIRONMENTAL RESOURCES LTD, Acid rain — a review of the phenomenon in the EEC and Europe, Graham and Trotman, London (1983).

[18] DAVIS, S.C., The European year of the environment, J. R. Soc. Arts **135** 5373 (1987) 676–686.

[19] SHEEHAN, P., et al. (Eds), Appraisal of Tests to Predict the Environmental Behaviour of Chemicals, SCOPE Monogr. 25, Wiley, New York (1985).
[20] INTERNATIONAL ATOMIC ENERGY AGENCY, Agrochemical Residue-Biota Interactions in Soil and Aquatic Ecosystems (Proc. Panel Vienna, 1978), IAEA, Vienna (1980).
[21] INTERNATIONAL ATOMIC ENERGY AGENCY, Agrochemicals: Fate in Food and the Environmental (Proc. Symp. Rome, 1982), IAEA, Vienna (1982).
[22] INTERNATIONAL ATOMIC ENERGY AGENCY, Isotope Tracer-Aided Studies of Agrochemical–Biota Interactions in Soil and Water, IAEA-TECDOC-247, IAEA, Vienna (1981).
[23] INTERNATIONAL ATOMIC ENERGY AGENCY, Agrochemical–Biota Interactions in Soil and Water Using Nuclear Techniques, IAEA-TECDOC-283, IAEA, Vienna (1983).
[24] NIELSEN, D.R., MACDONALD, J.G. (Eds), Nitrogen in the Environment, Vol. 1, Academic Press, New York (1978).
[25] INTERNATIONAL ATOMIC ENERGY AGENCY, Soil Nitrogen as Fertilizer or Pollutant (Proc. Panel Piracicaba, 1978), IAEA, Vienna (1980).
[26] COUGHTREY, P.J., et al., Radionuclide Distribution and Transport in Terrestrial and Aquatic Ecosystems, Vols 1–6, Balkema, Rotterdam (1983–1985).
[27] UNITED NATIONS SCIENTIFIC COMMITTEE ON THE EFFECTS OF ATOMIC RADIATION, Sources and Effects of Ionizing Radiation, UN, New York (1977).
[28] UNITED NATIONS SCIENTIFIC COMMITTEE ON THE EFFECTS OF ATOMIC RADIATION, Genetic and Somatic Effects of Ionizing Radiation, UN, New York (1986).
[29] GUTHRIE, F.E., PERRY, J.J. (Eds), Introduction to Environmental Toxicology, Elsevier, Amsterdam and New York (1980).
[30] CHRISTENSEN, H.E., LUGINBYHL, B.S. (Eds), The Toxic Substances List, United States Department of Health, Education and Welfare, Washington, DC (1974).
[31] INTERNATIONAL COMMISSION ON RADIOLOGICAL PROTECTION, Limits for Intakes of Radionuclides by Workers, ICRP Publication No. 30, Annals of the ICRP **2** 3/4; **4** 3/4; **6** 2/3 (1981).
[32] FOOD AND AGRICULTURE ORGANIZATION OF THE UNITED NATIONS, Pesticide Residues in Food — 1985, Plant Production and Protection, Paper No. 68, FAO, Rome (1986).
[33] FRANCIS, A.J., "Low level radioactive wastes in subsurface soils", Soil Reclamation Processes (TATE, R.L., KLEIN, D., Eds), Marcel Dekker, New York (1985) 279–331.
[34] INTERNATIONAL ATOMIC ENERGY AGENCY, Pesticide Residues and Radioactive Substances in Food: A Comparative Study of the Problems, IAEA-TECDOC-144, IAEA, Vienna (1972); also in Environ. Qual. Saf. **3** (1974) 17–34.
[35] WORLD HEALTH ORGANIZATION, Pesticide Residues in Food, Technical Report Series No. 458, WHO, Geneva (1970).
[36] WORLD HEALTH ORGANIZATION, Pesticide Residues in Food, Technical Report Series No. 592, WHO, Geneva (1976).
[37] FOOD AND AGRICULTURE ORGANIZATION OF THE UNITED NATIONS, Codex Alimentarius Commission, Report of the 16th Session, FAO, Rome (1985).

[38] WINTERINGHAM, F.P.W., Working Paper for 23rd Session of UNSCEAR, United Nations Scientific Committee on the Effects of Atomic Radiation, Vienna, 1974, 1-8.
[39] WINTERINGHAM, F.P.W., Foreign chemicals and radioactive residues in the biosphere — introduction, Chemosphere 2 2 (1973) 37-40 (and later publications in the series).
[40] Summary on 2,4-D, Chemosphere 2 2 (1973) 55-56.
[41] ALTMANN, H., WINTERINGHAM, F.P.W., Summary on tritium, Chemosphere 3 6 (1974) 263-264.
[42] Summary on pesticides — all, as a group, Chemosphere 2 2 (1973) 53-54.
[43] Summary on radionuclides — all, as a group, Chemosphere 2 5 (1973) 207-208.
[44] SAINSBURY, M , Natural products in the fight against cancer, Chem. Br. 15 3 (1979) 127-130.
[45] COX, J., Press report on longevity in the UK, The Times, London (21 Jan. 1987) 17.
[46] FARMER, P.B., Monitoring for human exposure to carcinogens, Chem. Br. 18 11 (1982) 790-794.
[47] WRIGHT, P., Press quote of the National Radiological Protection Board, Harwell, Report prepared on behalf of the Commission of the European Communities, The Times, London (25 Mar. 1987) 2.
[48] KAPRIO, L.A., Death on the Road, World Health Organization, Geneva (Oct. 1975) 4-9.
[49] WILSON, R., et al., Risk assessment and comparisons, Science 236 4799 (1987) 267-300.
[50] DOLL, R., Environmental chemicals and cancer, Chem. Br. 23 9 (1987) 847-850.
[51] BIXLER, G., SHEMILT, L.W. (Eds), Chemistry and World Food Supplies: The New Frontier, Vol. 1, Proceedings; Vol. 2, Perspectives and Recommendations, Pergamon Press, Oxford (1983).
[52] GIBBS, M., CARLSON, C. (Eds), Crop Productivity — Research Imperatives Revisited, Stage 1, Report; Stage 2, Report, Michigan State University, East Lansing (1986).
[53] WINTERINGHAM, F.P.W., "Soil nitrogen with particular reference to herbicidal and related chemical action", Proc. Symp. Pisa, 1979, Publication AC/4/65-83, Consiglio Nazionale delle Ricerche, Pisa (1979).
[54] COOK, L.M., WOOD, R.J., Genetic effects of pollutants, Biologist 23 3 (1976) 129-139.
[55] LORENZ, H., Binding forms of toxic heavy metals, mechanisms of entrance of heavy metals into the food chain, and possible measures to reduce levels in foodstuff, Ecotoxicol. Environ. Saf. 3 1 (1979) 47-58.
[56] HANCE, R.J., Herbicides and the soil, Chem. Br. 16 3 (1980) 128-156.
[57] KORTE, F., et al., Ecotoxicological profile analysis. Part I. Chemosphere 7 1 (1978) 79-102 (and subsequent parts in Ecotoxicol. Environ. Saf.).
[58] KORTE, F., Ecological Chemistry — State of the Art, Doc. 0-820, Gesellschaft für Strahlen- und Umweltforschung mbH, Munich (1984).
[59] WINTERINGHAM, F.P.W., Comparative ecotoxicology of halogenated hydrocarbon residues, Ecotoxicol. Environ. Saf. 1 3 (1977) 407-425.
[60] WINTERINGHAM, F.P.W., Agro-ecosystem — chemical interactions and trends, Ecotoxicol. Environ. Saf. 3 2 (1979) 219-235.

[61] COTTENIE, A., et al., Environmental Chemicals: Criteria for the Protection of Non-Human Biota in the Context of Agriculture, Forestry, Fisheries and Feed, Monograph Report prepared as FAO/UNEP Project 0107-76-01, Food and Agriculture Organization of the United Nations, Rome (1977).

[62] JURY, W.A., et al., Transport and transformations of organic chemicals in the soil–air–water ecosystem, Rev. Environ. Contam. Toxicol. **99** (1987) 119–164.

[63] JURY, W.A., Personal communication and demonstrations, University of California, Riverside, Nov. 1987.

IAEA-SM-297/45

Invited Paper

PESTICIDE DISSIPATION IN SOILS AS A MODEL FOR XENOBIOTIC BEHAVIOUR

J.B. WEBER
Department of Crop Science,
School of Agriculture and Life Sciences,
North Carolina State University,
Raleigh, North Carolina,
United States of America

Abstract

PESTICIDE DISSIPATION IN SOILS AS A MODEL FOR XENOBIOTIC BEHAVIOUR.
For the older 'hard' pesticides such as DDT dissipation meant primarily redistribution and bioaccumulation. For the majority of the pesticides currently being used, dissipation refers to all the degradation and transfer processes involved in the ultimate fate of chemicals in the environment. The degradation processes include non-biological, chemical and photochemical decomposition and biological decomposition by organisms and microorganisms. The transfer processes include pesticide sorption by soil colloids; diffusion of volatile pesticides into the atmosphere; movement downwards of pesticides into the soil in percolating waters, movement upwards with capillary water, and movement off the soil surface into surface waters; and absorption, exudation and bioaccumulation of pesticides by organisms. Knowledge of key pesticide and soil properties is necessary in understanding the fate and behaviour of pesticides. Key pesticide properties are symbolized by the acronym SILVER, which stands for solubility in water, ionizability, longevity (persistence), extractability in organic solvents, and the reactive groups present. The key soil properties are symbolized by the acronym SCOOP, which stands for structure (such as fragipans or macropores), clay type and amount, organic matter content, oxide (hydrous) content, and the pH of the system. These properties have been used in computer models to predict pesticide dissipation in soil and the environment. The greatest weakness of the models has been the inability to simulate real world conditions.

1. INTRODUCTION

The dissipation of pesticides in the environment has come to mean different things to different people. By definition, the word 'dissipation' has been defined as "the act of scattering, the condition of being scattered; dispersion; and wasteful consumption or expenditure". Rachel Carson [1], in her book, wrote of the "distribution" and "redistribution" of pesticides in the world. She stated that "Chemicals sprayed on croplands or forests or gardens lie long in the soil, entering into living organisms, passing from one to another in a chain of poisoning and death. Or, they pass mysteriously by underground streams until they emerge and, through the alchemy of air and sunlight, combine into new forms that kill vegetation, sicken cattle, and work unknown harm on those who drink from once pure wells". She implied that the chemicals do not really disappear, that they are merely transferred from soil to organisms and then to higher animals. Frank Graham [2], in his book, discussed the "disappearance" of pesticides, using George J. Wallace and Ernest A. Boykin's definition, which was "Since soil residues decline, and accretions in earthworms and presumable other soil organisms do not, we suggest that much of the so-called 'disappearance' of persistent chemicals is really transfer and re-distribution from soil to soil organisms, and then to higher animals". James Whorton [3], in his book, wrote of the disappearance of arsenic insecticides from soil by drainage before the next season "thus posing no threat to agriculture". "Weathering" was the term used to explain the disappearance of foliage applied arsenic insecticides from the fruit of treated fruit trees and vines. It referred primarily to the washing off of arsenic from the fruit by rainfall. Marco et al. [4], in their book, discussed the environmental fate of pesticides and compared the relative persistence of the old chlorinated hydrocarbon pesticides with the newer 'short lived' chemicals. Disappearance of pesticides included transport, transfer and degradation processes. It is apparent that in discussions of the dissipation of the older 'hard' or 'persistent' pesticides reference was made primarily to the scattering or distribution of the parent compounds in the environment by way of transfer and bioaccumulation and biomagnification processes. With regard to the dissipation of the more recently developed 'short-lived' pesticides, reference has been made to transfer and degradation processes.

Brown [5], in an ecological discussion of the disappearance of pesticides in the environment, discussed bioaccumulation, degradation and translocation of the compounds. Weber and colleagues [6, 7] discussed the fate and behaviour of herbicides in the environment in terms of degradation and transfer processes. In this paper the dissipation of pesticides is discussed primarily in terms of degradation and transfer processes, since chemicals which are found to bioaccumulate and biomagnify in living organisms in significant amounts will not be registered and thus sold commercially.

2. PESTICIDE DISSIPATION PROCESSES

Figure 1 illustrates the many processes involved in the dissipation of herbicides (pesticides) in the environment. The degradation processes are characterized by the splitting of the herbicide (HB) molecule. They include degradation by chemical, photochemical or biological decomposition processes. The transfer processes are characterized by the HB molecules remaining intact. They include sorption to soil

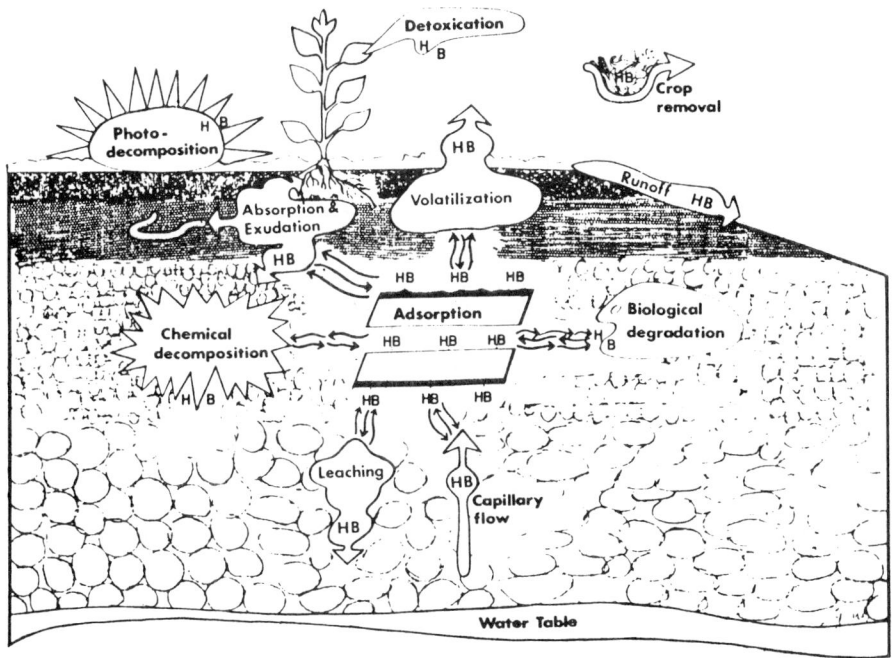

FIG. 1. Processes influencing the behaviour and fate of herbicides in the environment. Degradation processes are characterized by the splitting of the herbicide (HB) molecule. Transfer processes are characterized by the HB molecules remaining intact [6].

constituents, absorption, exudation and retention by plants and other organisms, movement off the soil in runoff and washoff, movement into the air by volatilization, movement downward in the soil in leachate, and upward in the soil through capillary action. Studies have been carried out using model terrestrial systems and a balance sheet approach [8, 9] to quantify the predominant processes involved in the dissipation of specific chemicals. Figure 2 depicts an apparatus used to quantify some of the processes involved in the dissipation of some s-triazine herbicides [8]. Table I shows the effect of the soil pH on the dissipation of atrazine and prometryn over a 5 month period using the apparatus depicted in Fig. 1 and a balance sheet approach [8]. At the end of the five month period, the majority of each herbicide was found in the surface 0 to 7.5 cm soil zone as parent compound or metabolite(s). Much smaller amounts were found in the subsurface 7.5 to 26 cm soil zone. Small amounts of the chemicals were taken up by crop plants and very small amounts were volatilized from the soil and/or lost from the soil in the leachate. Liming the soil was found to have a pronounced effect on the degradation, plant uptake (bioavailability) and movement of each herbicide into the subsurface soil and leachate.

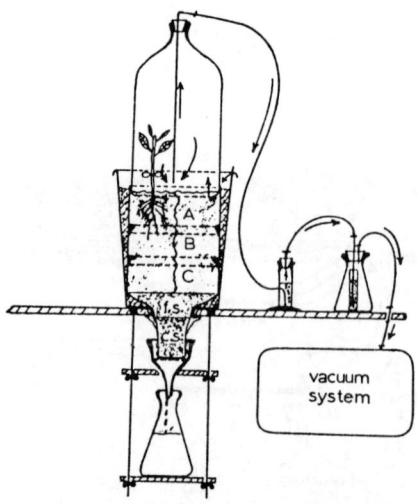

FIG. 2. *Model terrestrial ecosystem used in pesticide dissipation studies [8].*

Many books have been published which have addressed the various processes involved in the fate and behaviour of pesticides in the environment. The American Chemistry Society [10] and the Soil Science Society of America [11] were among the first scientific societies to publish symposia specifically directed towards this end. Both societies arranged for follow-up symposia which addressed the dissipation of pesticides and other contaminants in the environment [12–14]. Several books, including those by Goring and Hamaker [15], Hance [16], Khan [17] and Morrill et al. [18], have addressed the dissipation of pesticides and other organic pollutants specifically in soils. Some, including Hartley and Graham-Bryce's book [19], addressed the principles of pesticide behaviour and some, such as that of Camper [20], described and discussed the methods used to carry out pesticide dissipation studies. Others, such as Hill and Wright's book [21], addressed the microbial aspects of pesticide behaviour. Still others, such as Audus' [22] and Kearney and Kaufman's [23] books, addressed specifically the behaviour of herbicides in plants and soils. The most recent activity concerning pesticide and other organic pollutant dissipation in the environment has been the attempt to use models to simulate and predict pesticide fate and behaviour [24, 25]. Thus, over the past 25 years we have progressed to the point that we recognize that pesticides and many other chemicals that are added to the environment are dissipated by many different processes and that the ultimate disposition of each chemical is regulated by the properties of the parent compound and its metabolites, the properties of the soil, the climatic conditions and the relative interaction of each of the processes involved.

TABLE I. EFFECT OF pH ON THE DISSIPATION OF SOIL APPLIED ATRAZINE AND PROMETRYN USING A BALANCE SHEET APPROACH IN A MODEL TERRESTRIAL SYSTEM (^{14}C STUDIES)
(modified from Ref. [8])

Distribution	Atrazine		Prometryn	
	pH 5.5	pH 7.7[a]	pH 5.5	pH 7.5[a]
	(% of applied)[b]			
Volatilization				
Parent	0	0	0	0
CO_2	0.9	1.0	0.1	0.6
Plant uptake	1.6	4.3	0.6	1.6
Leachate	0.1	0.2	0.1	2.9
Soil (MeOH extractable)				
Surface 7.5 cm (A)				
Parent	12.7	38.3	47.0	11.3
Hydroxylated	36.9	10.9	26.8	19.8
Dealkylated	1.4	5.0	1.1	0.4
Residual	3.6	3.1	1.3	6.7
Unknown metabolite			2.6	6.2
Acetic acid extractable	25.6	19.8	14.1	18.1
Fixed	15.4	16.3	4.6	23.0
Subsurface 7.5–26 cm (B, C)	1.8	6.1	1.7	9.4
Total	100.0	100.0	100.0	100.0
Actual recovery	89.4	90.7	88.8	86.6

[a] Limed.

[b] Normalized to 100% recovered.

3. PESTICIDE PROPERTIES

When a formulated pesticide reaches the soil environment it is acted upon by many forces, especially water and soil particulate matter, causing the parent compound to be chromatographically separated from the adjuvants and/or the solubilizing salts, if the compound is an ionizable one. This is normally the case unless the pesticide has been formulated into a controlled release form that prevents immediate contact with the soil–water and particulate matter. Encapsulation, or some other type of controlled release formulation, delays the release of the parent chemical to be acted upon by the soil, but the same forces eventually come into play. Pesticides behave differently in soil, depending on their chemical, physical and biological properties. Classical chemical principles apply, depending on the ionizability, water solubility, volatility and degradability of a given compound. In teaching these principles to students it is useful to use the acronym SILVER to denote the key properties of a pesticide. 'S' represents the water solubility of a chemical at a specific temperature (normally 25°C), pressure (normally atmospheric pressure) and pH (normally 7). 'I' denotes how the chemical ionizes, i.e. whether it has acid or base properties and hence is active as an anion or cation, or is non-ionic and hence has no specific charge properties. 'L' signifies the longevity or persistence of a chemical under normal use conditions (normally expressed in weeks). 'V' represents the volatility of a chemical as expressed by its vapour pressure under specified conditions (normally 25°C and atmospheric pressure). 'E' represents the extractability of a pesticide in non-polar organic solvents such as octanol and is an expression of the lipophilic character of a compound. 'R' stands for the reactive groups present and denotes whether or not a pesticide has PO_3^- or AsO_3^- groups which can complex with soil clay minerals, or NO_2 groups which can H-bond to proteinaceous substances. Usually two or three of the properties are pre-eminent in regulating the fate and behaviour of a specific pesticide in the environment. Cationic pesticides, pesticides of extremely low water solubility and pesticides which readily complex with soil colloids are strongly bound to soil colloids and tend to be immobile in soils, while those that are anionic, highly water soluble and which do not possess reactive groups are highly mobile. Pesticides which have high vapour pressures must normally be incorporated or injected into the soil in order to persist long enough to effectively control the target pest. Pesticides which have low water solubilities are nearly always readily extractable into octanol or other non-polar solvents and are readily found associated with lipophilic particulate matter in soils.

4. SOIL PROPERTIES

Selected properties of the soil dictate to a great extent how a pesticide behaves, i.e. whether or not a chemical is adsorbed, is mobile or immobile, or is or is not biologically available. SCOOP is a useful acronym for denoting the key properties of the soil that regulate pesticide behaviour. 'S' represents structure, which denotes the presence of fragipans or hardpans that would impede the infiltration of water and/or pesticides downwards into the soil, or the presence of macropores, resulting from the decay of dead roots or from animal activities, which would greatly accelerate water and pesticide movement into the soil. 'C' denotes the type and amount of clay in the soil, which might react with pesticides and impede their downward movement or their capillary movement back towards the surface or, in the case of expanding type clays, may interfere with their biological availability, resulting in persistence or 'carry-over' problems. 'O' represents the organic matter component of the soil that readily adsorbs pesticides and decreases their movement in the soil and their bioavailability. The second 'O' represents the hydrous iron and aluminium oxide content of the soil that may bind and reduce the movement and bioavailability of the anionic species of acid pesticides. 'P' denotes the soil pH, an important property that regulates the predominant ionic species of ionizing pesticides in solution and the predominant microbial species which are present.

5. PESTICIDE DISSIPATION AS A XENOBIOTIC MODEL

To understand and predict the fate and behaviour of pesticides in soils, investigators utilize the properties of the pesticides and the soils in conjunction with the predominant processes that occur when pesticides are applied to soil systems.

5.1. Classification of pesticides

Pesticides are classified according to their ionizing characteristics as strong, moderate or weak acids or bases, or non-ionic compounds; water solubility characteristics as high, moderate or low in water solubility; volatility characteristics as high, moderate or low in volatility; extractability in octanol characteristics as being high, moderate or low in lipophilicity; and soil reactivity characteristics according to whether or not they possess PO_3^-, AsO_3^- or NO_2 groups (Table II).

5.2. Sorption to particulate matter

Pesticides become bound to soil colloids through a variety of adsorption mechanisms, including ionic bonding; physical bonding by way of H-bonds, charge–transfer complexes, or hydrophobic bonds; ligand exchange; and/or precipitation reactions. These adsorption mechanisms express themselves in the form of different types of adsorption isotherms, as shown in Fig. 3. Strongly basic pesticides such as diquat are bound to organic matter and clay minerals through strong ionic bonds [26]. Weakly basic pesticides such as prometryn are adsorbed through ionic bonds and/or by physical adsorption, depending on the pH of the system. Acidic pesticides such as dicamba are adsorbed through physical adsorption when in the molecular form and repelled or adsorbed in low amounts when in the anionic form, depending on the pH of the system and the pK_A of the compound. Non-ionic pesticides such as chlorpropham are adsorbed by physical adsorption and the amount of adsorption is generally inversely related to the water solubility of the compound [27].

TABLE II. CLASSIFICATION SCHEME FOR PESTIDICES ACCORDING TO SILVER (solubility in water, ionizability, longevity, volatility, extractability in octanol and reactive groups present)

Property	Class, description	Parameter
Water solubility		Log water solubility (ppm)
	Very highly soluble	>4
	Highly soluble	3 to 4
	Moderately soluble	2 to 3
	Low solubility	1 to 2
	Very low solubility	0 to 1
	Extremely low solubility	<0
Ionizability		pK_A
	Acid, very strong	>1
	Acid, strong	1 to 3
	Acid, moderate	3 to 5
	Acid, weak	5 to 7
	Acid, very weak	7 to 9
	Acid, extremely weak	<9
	Base, extremely weak	>1
	Base, very weak	1 to 3
	Base, weak	3 to 5
	Base, moderate	5 to 7
	Base, strong	7 to 9
	Base, very strong	<9
Longevity		Longevity (weeks)
	Very short lived	<4
	Short	4 to 20
	Moderate	20 to 40
	Long	40 to 60
	Very long	60 to 80
	Extremely long	>80
Volatility		Log vapour pressure (mm Hg)
	Very high	>−3
	High	−4 to −3
	Moderate	−5 to −4
	Low	−6 to −5
	Very low	−7 to −6
	Extremely low	<−7

TABLE II (cont.)

Property	Class, description	Parameter
Octanol extractability		Log K_{oc}
	Extremely lipophilic	>5
	Very highly lipophilic	4–5
	Highly lipophilic	3–4
	Moderately lipophilic	2–3
	Low lipophilicity	1–2
	Very low lipophilicity	<1
Reactive groups		*Reactivity*
	PO_3^- groups present	High
	AsO_3^- groups present	High
	NO_2 groups present	Moderate
	No reactive groups present	None

Pesticides which contain NO_2 groups such as trifluralin are adsorbed to proteinaceous organic colloids through H-bonds, while those which contain PO_3^- groups such as glyphosate are bound to clay minerals through ligand exchange. Adsorption of cationic species produces H type isotherms, while adsorption of molecular species produces L, S or C type isotherms [7]. Adsorption of mixtures of cationic and molecular species produces L type isotherms, which are well described by the Langmuir equation [28]. Adsorption of pesticides which exhibit L type isotherms are relatively well described by the Freundlich equation [29]. The relative sorption of non-ionic pesticides by soils can be evaluated by a comparison of the Freundlich K and 1/n constants for each pesticide [15]. As a general rule, strongly basic pesticides, pesticides which possess reactive groups and pesticides of extremely low water solubility are bound by soils in high amounts (Table II, Fig. 3). Weakly basic pesticides are adsorbed by soils in moderate amounts and weakly acidic pesticides are adsorbed in very low amounts. Adsorption of both acidic and basic pesticides is inversely related to the soil pH.

5.3. Mobility of pesticides over and in soils

Pesticides may be carried off soils and into surface waters in one of two ways: either dissolved in the runoff water or adsorbed on to particulate matter that is lost in erosion processes. Wauchope [30] reviewed the losses of pesticides from agricultural fields and stated that the amounts lost ranged from 5 to 16%. The highest losses

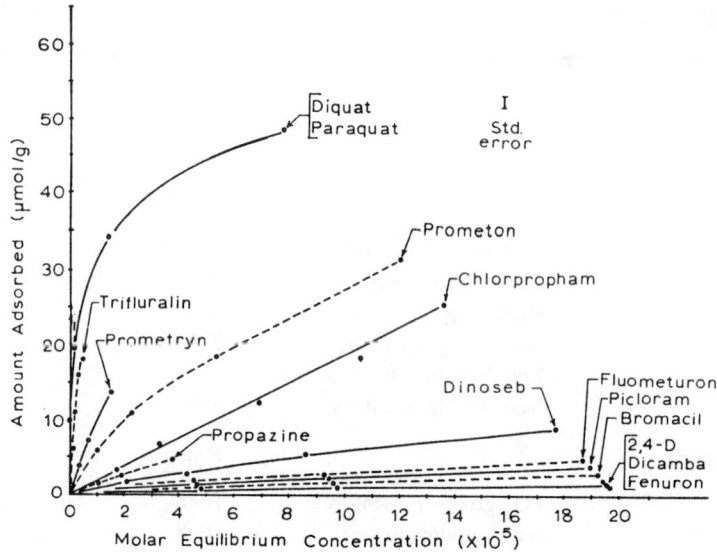

FIG. 3. *Isotherms for adsorption of 14 herbicides by soil organic matter [26].*

occurred from wettable powder formulations that had been applied to the soil surface and followed shortly after application by intense rains. Weber et al. [31], in an evaluation of long term pesticide losses from agricultural fields, reported that losses were generally in the 0.5 to 2.2% range and that the overall mean loss was 1.2%. Losses were lowest from pastures and grassed areas and where conservation tillage practices were utilized in crop production.

Pesticides move in the soil by two major processes: diffusion and mass flow [32]. Volatile pesticides diffuse through and out of the soil in the vapour phase. Losses are generally related to the vapour pressure of the pesticide and are higher from warm, moist soils than they are from cool, dry soils. Pesticides with high vapour pressures must be incorporated into the soil to decrease the vapour losses and allow the chemical to persist long enough to control the target pest. Vapour losses of pesticides from soils range from nearly 100% for highly volatile fumigants such as methyl bromide to very low levels for soil incorporated volatile herbicides such as trifluralin.

Pesticides dissolved in soil solution are carried into the soil by the process of mass flow. They normally adsorb and desorb to soil particulate matter as percolating water carries them in a downward direction, and they return to the soil surface when capillary water carries them upwards during periods of high evaporation. Movement of pesticides in the soil is normally inversely related to pesticide adsorption by the soil particulate matter. Movement of pesticides in soil may also occur when percolating water passes through macropores resulting from decaying roots and

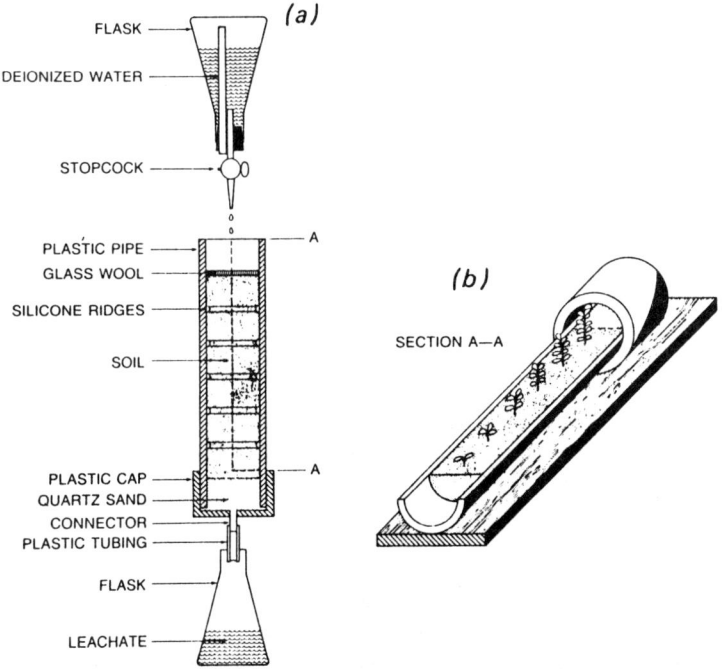

FIG. 4. Soil pesticide leaching column set up for unsaturated flow (a) and for plant bioassay of split column after leaching has been completed (b) [33].

animal burrows, or through cracks resulting from the shrinkage of expanding type clay minerals. Water and pesticide movement may also be impeded by the presence in the soil of fragipans or hardpans. The relative mobility of pesticides in soil is measured in the laboratory by use of soil leaching columns [33], soil thin layer chromatographic plates [34], and soil thick layer chromatographic trays [35]. The methods produce relative information. Figure 4 shows a soil leaching column set up for unsaturated flow pesticide mobility studies (a) and the split column (b) arranged for bioassaying the herbicide after leaching is completed [33]. The relative mobility of selected pesticides in soil column studies is evaluated by a comparison of the cumulative pesticides appearing in the leachate, as shown in Fig. 5, and a comparison of the distribution of the pesticides remaining in the soil column, as shown in Table III [36]. Analysis of the leachate showed that bromacil was slightly more mobile than buthidazole and that both herbicides were much more mobile than prometon, atrazine or diuron (Fig. 5). It also showed that bromacil and buthidazole began to appear in the leachate after 12 cm of water had been applied and that losses of the herbicides reached 5.2 and 1.9%, respectively. A comparison of the herbicide distribution in the soil after leaching showed that the moderately soluble, extremely weak acid bromacil was slightly more mobile than the highly soluble, amphoteric

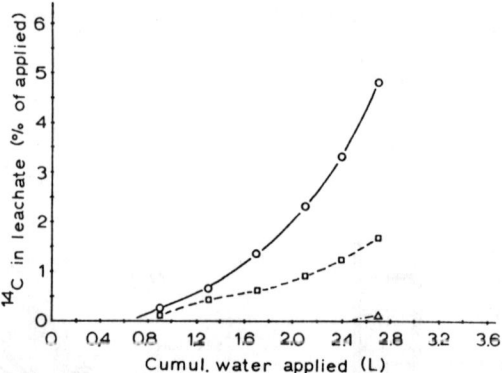

FIG. 5. Cumulative pesticide in leachate from Lakeland sandy loam soil treated with bromacil (O), buthidazole (□), and prometon, atrazine and diuron (△), and leached with 1.2 cm/d for 30 days [36].

buthidazole, followed by the low solubility, very weak base atrazine, the moderately soluble, weak base prometon, and the low solubility, non-ionic diuron (Table III). Descriptions of the properties of the herbicides were made according to the classification scheme shown in Table II.

Capillary movement studies of pesticides indicate that highly mobile pesticides such as imazaquin [37] and chlorsulfuron [38] do readily move upwards in capillary water from subsoil to the surface, as well as downwards in percolating water. This is especially the case where free water, such as a high water table or a low wet area in a field, is present to maintain capillary flow in an upward direction over long periods of time. More than 20% of the leached pesticides were returned to the soil surface when half of the applied water was allowed to evaporate, suggesting that the capillary component of pesticide mobility is important in modelling pesticide movement in soils.

5.4. Degradation of pesticides

Many references are available which address the degradation of pesticides in soils, including those of Hill and Wright [21], Kearney and Kaufman [23], Haque and Freed [39], and Saltzmann and Yaron [40]. Essentially, a selected pesticide is either degraded chemically, photochemically or biologically, or by one or more of these processes. Longevity studies carried out in the field, at several locations under diverse soil and climatic conditions, provide longevity values usually in weeks of biological activity [41]. Table II provides a classification scheme for the relative longevity or life expectancy of pesticides in the soil. Laboratory studies are carried out to quantify the relative importance of photochemical, chemical and microbial

TABLE III. DISTRIBUTION IN LAKELAND SANDY LOAM SOIL TREATED WITH ^{14}C LABELLED HERBICIDES AND LEACHED WITH 1.2 cm/d FOR 30 DAYS (modified from Ref. [36])

Soil depth (cm)	Bromacil	Buthidazole	Atrazine (% of applied)[a]	Prometon	Diuron
0–5	8.0	19.6	62.3	96.3	99.6
5–10	14.4	17.8	22.1	3.2	0.3
10–15	20.2	18.5	14.8	0.4	0.4
15–20	22.3	18.7	0.7	0.0	0.0
20–25	18.6	15.1	0.0	0.0	0.0
25–30	11.3	8.3	0.0	0.0	0.0
Total	94.8	98.1	99.9	99.9	99.9
In leachate	5.2	1.9	0.1	0.1	0.1

[a] Percentage recovered for each herbicide was 91.6, 83.5, 114.3, 87.5 and 107.7, respectively. All values were normalized to 100% recovered.

degradation for a given pesticide and to allow for the identification of the degradation pathways and the organisms responsible. Half-life values are obtained for each process and, taken together with the longevity values obtained from field studies, provide an indication of the length of time that a given pesticide remains intact in the environment.

6. SUMMARY

The dissipation of pesticides in the soil involves both degradation and transfer processes. Degradation may be by chemical, photochemical or biological processes. Transfer processes include absorption, exudation and retention of the unaltered pesticide by plants or other organisms; retention and release by soil particulate matter; movement of pesticide vapours from the soil to the atmosphere; movement of pesticides downwards through the soil in percolating water, upwards through the soil in capillary water, and movement off the soil into surface waters. The behaviour and fate of a given pesticide is regulated by its chemical, physical and biological properties, the properties of the soil and the climatic conditions. Use of these parameters is invaluable in predicting and modelling the dissipation of pesticides, provided the real world can be simulated to an acceptable degree.

ACKNOWLEDGEMENTS

The author acknowledges the Gesellschaft für Strahlen- und Umweltforschung mbH, Munich, for providing him with the opportunity of participating in this symposium and E. Stewart for helping in the preparation of this manuscript at very short notice.

REFERENCES

[1] CARSON, R., Silent Spring, Houghton Mifflin, Boston (1962) 368.
[2] GRAHAM, F., Since Silent Spring, Houghton Mifflin, Boston (1970) 333.
[3] WHORTON, J., Before Silent Spring, Princeton University Press, New Jersey (1974) 289.
[4] Silent Spring Revisited (MARCO, G.J., HOLLINGWORTH, R.M., DURHAM, W., Eds), American Chemical Society, Washington, DC (1987) 214.
[5] BROWN, A.W.A., Ecology of Pesticides, Wiley, New York (1978) 525.
[6] WEBER, J.B., MONACO, T.J., WORSHAM, A.D., What happens to herbicides in the environment? Weeds Today **4** 1 (1973) 16.
[7] WEBER, J.B., WEED, S.B., "Effects of soil on the biological activity of pesticides", Pesticides in Soil and Water (GUENZI, W.D., Ed.), Soil Science Society of America, Madison, WI (1974) 223.
[8] BEST, J.A., WEBER, J.B., Disappearance of s-triazines as affected by soil pH using a balance sheet approach, Weed Sci. **22** (1974) 364.
[9] WEBER, J.B., "Soils, herbicide sorption, and model plant-soil systems", Research Methods in Weed Science (CAMPER, N.D., Ed.), Southern Weed Science Society, Champaign, IL (1986) 155.
[10] Organic Pesticides in the Environment (GOULD, R.F., Ed.), American Chemical Society, Washington, DC (1966) 309.
[11] Pesticides and their Effects on Soils and Water (BRETH, S.A., STELLY, M., Eds), Soil Science Society of America, Madison, WI (1966) 150.
[12] Fate of Organic Pesticides in the Aquatic Environment (GOULD, R.F., Ed.), American Chemical Society, Washington, DC (1972) 280.
[13] Pesticides in Soil and Water (GUENZI, W.D., Ed.), Soil Science Society of America, Madison, WI (1974) 562.
[14] Residues of Pesticides and Other Contaminants in the Total Environment, Residue Reviews (GUNTHER, F.A., Ed.), Vol. 85, Springler-Verlag, New York (1983) 307.
[15] Organic Chemicals in the Soil Environment (GORING, C.A.I., HAMAKER, J.W., Eds) (2 Vols), Marcel Dekker, New York (1972) 968.
[16] Interactions Between Herbicides and the Soil (HANCE, R.J., Ed.), Academic Press, London (1980) 349.
[17] KAHN, S.U., Pesticides in the Soil Environment, Elsevier, Amsterdam (1980) 240.
[18] MORRILL, L.G., MAHILUM, B.C., MOHIUDDIN, S.H., Organic Compounds in Soils: Sorption, Degradation and Persistence, Ann Arbor Science, Ann Arbor, MI (1982) 326.

[19] HARTLEY, G.S., GRAHAM-BRYCE, I.J., Physical Principles of Pesticide Behaviour (2 Vols), Academic Press, London (1980) 1024.
[20] Research Methods in Weed Science (CAMPER, N.D., Ed.), Southern Weed Science Society, Champaign, IL (1986) 486.
[21] Pesticide Microbiology (HILL, I.R., WRIGHT, S.J.L., Eds), Academic Press, London (1978) 844.
[22] Herbicides — Physiology, Biochemistry, Ecology (AUDUS, L.J., Ed.) (2 Vols), Academic Press, London (1976) 564 (Vol. 1); 608 (Vol. 2).
[23] Herbicides — Chemistry, Degradation and Mode of Action (KEARNEY, P.C., KAUFMAN, D.D., Eds) (2 Vols), Marcel Dekker, New York (1976) 1036.
[24] Fate of Chemicals in the Environment (SWANN, R.L., ESCHENROEDER, A., Eds), American Chemical Society, Washington, DC (1983) 320.
[25] Vadose Zone Modeling of Organic Pollutants (HERN, S.C., MELANCON, S.M., Eds), Lewis Publishers, Chelsea, MI (1986) 295.
[26] WEBER, J.B., "Interactions of organic pesticides with particulate matter in aquatic and soil systems", Fate of Organic Pesticides in the Aquatic Environment (GOULD, R.F., Ed.), American Chemical Society, Washington, DC (1972) 55.
[27] CARRINGER, R.D., WEBER, J.B., MONACO, T.J., Adsorption–desorption of selected pesticides by organic matter and montmorillonite, J. Agric. Food Chem. **23** (1975) 568.
[28] WEBER, J.B., "Mechanisms of adsorption of *s*-triazines by clay colloids and factors affecting plant availability", The Triazine Herbicides (GUNTHER, F.A., Ed.), Springer-Verlag, New York (1970) 93.
[29] KOZAK, J., WEBER, J.B., Adsorption of five phenylurea herbicides by selected soils of Czechoslovakia, Weed Sci. **31** (1983) 368.
[30] WAUCHOPE, R.D., The pesticide content of surface water draining from agricultural fields — a review, J. Environ. Qual. **7** (1978) 459.
[31] WEBER, J.B., SHEA, P.J., STREK, H.J., "An evaluation of nonpoint sources of pesticide pollution in runoff", Environmental Impact of Nonpoint Source Pollution (OVERCASH, M.R., DAVIDSON, J.M., Eds), Ann Arbor Science, Ann Arbor, MI (1980) 69.
[32] LETEY, J., FARMER, W.J., "Movement of pesticides in soil", Pesticides in Soil and Water (GUENZI, W.D., Ed.), Soil Science Society of America, Madison, WI (1974) 67.
[33] WEBER, J.B., SWAIN, L.R., STREK, H.J., SARTORI, J.L., "Herbicide mobility in soil leaching columns", Research Methods in Weed Science (CAMPER, N.D., Ed.), Southern Weed Science Society, Champaign, IL (1986) 189.
[34] HELLING, C.S., TURNER, B.C., Pesticide mobility: Determination by soil thin-layer chromatography, Science **172** (1968) 562.
[35] GERBER, H.R., ZIEGLER, P., DUBACH, P., Leaching as a tool in the evaluation of herbicides, Proc. Br. Weed Contr. Conf. **10** (1970) 118.
[36] WEBER, J.B., WHITACRE, D.M., Mobility of herbicides in soil columns under saturated and saturated-flow conditions, Weed Sci. **30** (1982) 579.
[37] McKINNON, E.J., WEBER, J.B., unpublished data (1987).
[38] MAHNKEN, G.E., WEBER, J.B., unpublished data (1987).

[39] Environmental Dynamics of Pesticides (HAQUE, R., FREED, V.H., Eds), Plenum Press, New York (1975) 387.
[40] Pesticides in Soil (SALTZMAN, S., YARON, B., Eds), Van Nostrand Reinhold, New York (1986) 377.
[41] Herbicide Handbook, Weed Science Society of America, Champaign, IL (1983) 515.

IAEA-SM-297/35

Invited Paper

MOVEMENT OF PESTICIDES FROM THE SITE OF APPLICATION

J.R. PLIMMER
Environmental Chemistry Laboratory,
Agricultural Environmental Quality Institute,
Agricultural Research Service,
United States Department of Agriculture,
Beltsville, Maryland,
United States of America

Abstract

MOVEMENT OF PESTICIDES FROM THE SITE OF APPLICATION.
Although the types of pesticides that will be used in future years will change, reliance will continue to be placed on chemical pest control because the technology is well established and there is an expanding world market. Important among the data needed to reduce the environmental risks potentially associated with pesticide use is a clear understanding of environmental processes. Much information has been accumulated and this is discussed with special reference to the research being carried out by scientists of the Agricultural Research Service of the United States Department of Agriculture at Beltsville. Physical and chemical properties are important in predicting the environmental fate of a pesticide. Routes of dissipation from the site of application will be influenced by the agricultural system and the method of application. Causes of loss include runoff and a three year study showed the extent of herbicide runoff to the Wye River Estuary. Potential contamination of groundwater by leaching has been recognized as a problem in certain cases and its extent and causes are being extensively studied. Volatilization may be a significant source of loss. Two important factors are the vapour pressure of the pesticide and the nature of the surface. Pesticide volatilization was measured using specially designed sampling equipment. Losses from dry soil surfaces were much smaller than the rapid losses from moist surfaces. Losses were predictable in terms of the vapour pressure of the compound and its interaction with surfaces. Atmospheric movement is important in the transport of pesticides. They may be redeposited on the Earth's surface by a number of mechanisms including rainfall, fog and snow or be adsorbed on particulate matter. A specially constructed sampler was used to collect the liquid and vapour phases of fog. Pesticides, their alteration products, plasticizers, flame retardants, industrial chemicals and combustion products were found in fog samples collected in Maryland and California. Distribution between liquid and vapour phases was not in accordance with Henry's Law and experiments are in progress to measure Henry's Law constants to validate these observations.

1. INTRODUCTION

In December 1984, an FAO/IAEA Advisory Group met in Vienna to discuss the application of nuclear techniques to study environmental processes influencing pesticide behaviour [1]. The discussion focused on abiotic processes, both chemical and physical. Although biological processes were mentioned, this topic was intentionally not addressed in detail. The report contained guidelines for ecosystem research and there is no doubt that model ecosystems are an excellent route to information concerning the distribution and dissipation of pesticides within the ecosystem using ^{14}C labelled pesticides. Such information is essential for formulating guidelines for pesticide use, because of its value in predicting the environmental impact of a pesticide. When such information is integrated with the knowledge of toxicological effects of the parent compound and its alteration products there is an initial basis for recommendations concerning the judicious use of the compound in agriculture.

Although much information can be obtained by studies in the laboratory and in the field prior to official approval and widespread use of a pesticide, there must also be an adequate programme of surveillance and monitoring to identify effects on human health and the environment during the period of use of the pesticide. Regulatory or control systems for agrochemicals and other man made chemicals that enter the environment depend on the ability to make predictions from a knowledge of the physical and chemical properties of a substance, its toxicological effects and its interactions with many environmental components under a wide range of conditions.

Because similar information is needed for metabolites and alteration products, it is easy to realize that the problem of obtaining this information presents overwhelming difficulties. These problems are reflected in the dilemmas and delays of regulatory processes and on the increasing costs of obtaining data which must be borne by industry and ultimately by the consumer. The necessity for continued monitoring after the registration process is complete cannot be too heavily emphasized because in recent years it has brought to light some of the inadequacies in our understanding of the environmental behaviour of pesticides.

Pesticides may be applied to kill weeds or insects, or to control diseases or fungi. They are necessary to reduce crop losses that may amount to 45% of the total world food production. The pesticide market in 1985 was about US $13 800 million. It should increase to about US $15 700 million in the next 5 years and there is a growing market in the developing countries which used about 20% of the world's production in 1984 [2–4].

There has been an increased understanding of the implications of continued chemical pest control in terms of the environment, human health and pest management. This has resulted in the development of compounds that are less persistent in the environment, of lower mammalian toxicity, and frequently of greater selectivity towards pest species. The persistent organochlorine insecticides introduced in the late 1940s have been succeeded by the organophosphate and carbamate insecticides.

More recently the synthetic pyrethroids, growth regulators and other novel insecticides have been introduced. The range of herbicides available has widened and compounds now available, such as the sulphonyl-ureas, show activity at rates as low as 100 g/ha. Development of chemicals continues in order to overcome environmental problems and the problem of resistance that may develop in pest species. However, the increasing cost of introducing new chemicals to the market may now exceed US $40 million per chemical and this has considerably reduced the flow of new compounds. Because the world market for pesticides is very large, the stimulus to develop new compounds remains strong but we must expect in future to see additional developments based on biotechnology, microbial pesticides, behaviourally active compounds, and other new discoveries based on improved knowledge of pest biology. Better methods of application and formulation techniques will also be important.

These points concerning future development must be included here because we now understand more clearly the potential value and the limitations of chemical pest control where we have gained experience of pesticide use over an extended period of time. This paper will summarize some of the knowledge of environmental pathways and it will be based on research conducted primarily in the eastern United States by scientists of the Agricultural Research Service, United States Department of Agriculture.

Studies were undertaken at the Environmental Chemistry Laboratory of the Beltsville Agricultural Research Center over a period of more than 20 years to gain an understanding of the fate of pesticides applied in the field and the effects of different systems of agricultural management on the rates and pathways of pesticide loss. More recently, the research findings indicate the significance of the routes by which pesticides may be deposited on the Earth's surface.

The principal pathways by which pesticides may move from the site of application are shown in Fig. 1. The discussion in this paper will be limited to losses after application. There may often be high potential for losses to the environment during the process of application when evaporative or drift loss may be substantial depending on the formulation type, application equipment and type of chemical. The losses occurring after application will depend on climate, the type of active ingredient, formulation, the nature of the environmental surfaces upon which the chemical resides and on the characteristics of soil and plants at the target site. Investigation of these losses is important if we seek to reduce contamination of non-target sites and optimize the agricultural benefits of expensive chemical inputs.

2. FACTORS IMPORTANT IN DETERMINING ENVIRONMENTAL BEHAVIOUR

The factors that determine environmental behaviour of a compound are its chemical structure, its physical properties, the type of formulation, the method of

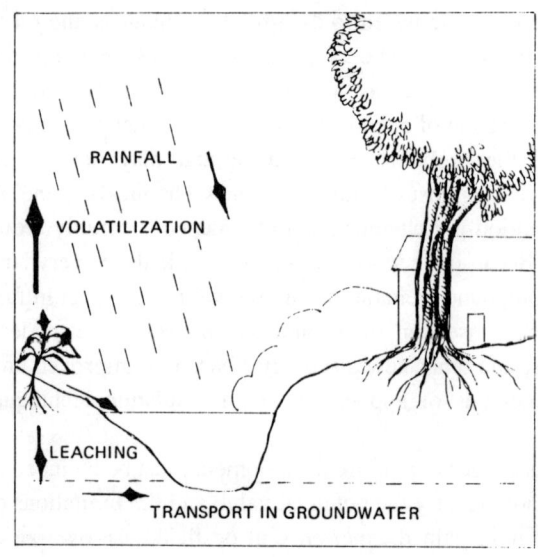

FIG. 1. Principal environmental pathways.

TABLE I. SELECTED PHYSICAL PROPERTIES OF PESTICIDES [5]

Pesticide	Vapour pressure at 20°C (Pa)	Solubility in water at 20°C (mg/L)	Log $K_{o/w}$
Atrazine	0.00004	30	2.4
Heptachlor	0.03	0.1	3.9
DDT	0.00002	0.003	6
Dieldrin	0.0005	0.17	3.7
Diazinon	0.008	38	3.3
Parathion	0.0006	15	3.8
Endosulfan	0.0011	0.15	3.6
Simazine	0.0000085	5	1.9
Trifluralin	0.006	0.5	3

$K_{o/w}$ = octanol–water partition coefficient.

application, and the local climatic and agricultural conditions. The biodegradability and biological activity are fundamentally linked to chemical stucture and much ingenuity has been invested in attempts to synthesize molecules which optimize both biological activity and an appropriate rate of degradation in the environment, as has been achieved in the case of the pyrethroid insecticides. The use of organochlorine insecticides has been restricted because they have come to be recognized as environmentally undesirable. They have low water solubility and they are quite stable. Because their lipid solubility is high, they tend to accumulate in the tissues of fish, birds, or mammals. Although their vapour pressures are low, they have entered the atmosphere and have become ubiquitous environmental contaminants. Fortunately, specific and sensitive methods are available for their analysis and this has facilitated studies of their environmental fate.

The physical properties that primarily influence pesticide behaviour in the environment are water solubility, vapour pressure, and the octanol–water partition coefficient. Some selected values are shown in Table I [5]. These values are useful indicators of potential volatility, soil mobility, bioaccumulation potential, etc.

3. AGRICULTURAL SYSTEMS

A pesticide may be applied to a soil surface or a standing crop. It is becoming an increasingly common agricultural practice in the USA to use some type of reduced tillage system, with no-till as the extreme. Crop residues are allowed to remain on the soil and subsequent planting takes place without disturbing these residues. The advantage of reduced tillage is the control of soil erosion and the reduction in labour costs. Herbicide application is used to control unwanted vegetation. There is high conservation of moisture and labour costs are minimized. During the decade preceding 1982, no-tillage acreage increased from 3.3 million to 10.5 million acres and minimum tillage rose from 26.3 to 95.5 million acres in the USA. Pesticides applied to these cultural systems will be intercepted by the mulch of plant material which covers the soil. This will retain a portion of the pesticide and it may be adsorbed by the organic matter, thus affecting volatilization and runoff. The mulch of plant material which covers the soil will be an excellent source of microbial nutrients and degradation rates may be enhanced.

4. ESTIMATES OF LOSS

It is difficult to obtain precise measurements of the quantitative aspects of pesticide dissipation. Some approximations may be useful. The most important routes are biological–chemical degradation and volatilization. The latter may account for 40 to 80% of the loss depending on the physical properties of the compound and its environment. Loss by runoff is probably no more than 5% and loss by leaching may account for less than 1%.

4.1. Runoff

Runoff depends on rainfall and the amount of runoff is linked to the time interval between rainfall and application. A study was conducted in the Chesapeake Bay area to measure atrazine runoff to the Wye River, a small tributary of the bay on the eastern shore of Maryland [6]. In this region, the continuous culture of corn is a common practice and atrazine is used in weed control in both no-till and conventional corn. Because corn is planted close to the river there was a concern that herbicide runoff might be associated with the decline of submerged aquatic vegetation which coincided with the introduction of atrazine use. Water and sediment were sampled at various points in the Wye River Channel as part of a three year programme to study this problem. Atrazine runoff was measured by determining the concentration of the herbicide at the edge of a 20 ha cornfield after application of atrazine and simazine at 1.68 kg/ha. The atrazine content was maximal in the first significant volume of runoff caused by rainfall during the first two weeks after application. The amount of atrazine in runoff declines significantly with time after application [7]. Atrazine moved primarily in the aqueous phase and movement effectively ceased 4 to 6 weeks after application. Because less simazine is used the quantity moving to the river was about one-tenth that of atrazine. Concentrations of atrazine at sites in the Wye River ranged from 0.9 μg/L at the mouth to about 15 μg/L at the source. Such low levels would not be expected to significantly inhibit the growth of submerged aquatic vegetation. The study indicated that atrazine runoff under present use patterns was unlikely to be a significant factor in the disappearance of submerged aquatic vegetation.

4.2. Leaching to groundwater

Pesticides have been found in well water as contaminants at many sites in the United States. Although the levels are usually below those which are toxicologically significant, the widespread occurrence of contamination has provoked much concern. In March 1987 [8], it was reported that 23 different pesticides or metabolites had been found in 24 states. There were detectable levels of pesticides in 11 000 of 45 000 wells examined and, of these, 5500 exceeded some health guidance level. The United States Environmental Protection Agency is conducting a national survey to determine the frequency of occurrence and concentration of pesticides in drinking water wells in a variety of settings, ranging from high pesticide use and high groundwater vulnerability to rare pesticide use and low groundwater vulnerability.

Although some contaminants may be leached from chemical spills, it has become clear that under certain conditions agricultural use of pesticides may be the source of contaminated water and several pesticides have been detected frequently in water samples, including atrazine, aldicarb and alachlor.

The flow of water through the porous structure of soils is normally slow and the processes of sorption and biodegradation take place during movement at such rates that pesticides are removed during the flow process and do not reach groundwater. However, alternative routes may be available by which pesticides may be transported more rapidly to groundwater. One mechanism by which pesticides may enter groundwater with the infiltrating water front involves rapid flow through channels (macropores) that could be produced by decay of organic matter, burrowing of invertebrates, desiccation, etc. Another mechanism is that of unusually rapid transport of pesticides as complexes formed by complexation with mobile colloids of clay or organic matter and it is feasible that some pesticides that are normally strongly adsorbed to soils could be transported in this way [9].

It is clear that leaching to groundwater is a pathway of pesticide loss. Although this is probably a very small fraction of that applied, it is an unforseen consequence of frequent pesticide use that demands further investigation and changes to remedy the situation. The question of macropores also challenges some assumptions concerning movement in soils and emphasizes a major problem in environmental studies, that of working in media which are heterogeneous and non-uniform. This raises difficulties when data obtained in laboratory or model experiments must be extrapolated to field conditions.

4.3. Volatilization

Many pesticides are applied as sprays using various types of equipment. There is considerable potential for loss during the spray operation when pesticides may be lost by spray drift or by evaporation. Losses may be reduced by improvements in formulation and application techniques. The present discussion will be concerned only with losses after application.

It may be considered that the applied pesticide is deposited as a thin layer of material that uniformly covers the treated surface of soil or foliage. The rate of evaporation from this film should be independent of the amount of material, i.e. 'zero order'. However, when material has evaporated, the surface is no longer uniformly covered and there are 'islands' of pesticide deposit. The rate of evaporation will then depend on the area covered and will be reduced as the area diminishes. The rate of loss should therefore become 'first order' as it will depend on the amount of material present. As residues decline to leave only those that reside in less accessible situations, there will again be changes in the kinetics of loss. Also important is the nature of the surface. The waxes or other lipids present on the leaf surface may adsorb organics and reduce rates of evaporation [10].

On a dry soil surface a spray will move into the soil by capillary action and the pesticide will become adsorbed. The rate of volatilization will be dependent on diffusion to the surface. In moist soils, evaporation of water will provide a driving force for movement of water to the surface and the pesticide will be transported in the soil solution to the surface where it will volatilize. Volatilization of lindane and dieldrin from a moist soil surface takes place up to five times faster than from a dry surface where the compounds are tightly adsorbed and the path to the surface is diffusion controlled [11].

The field situation is more complex. Rainfall affects movement of water into soil and movement of water to the surface will be reduced as the humidity increases. Water will move to the plant root zone and will be returned to the air by evapotranspiration which provides a continuous mechanism for removing water from soil. Evapotranspiration also ensures that leaf surfaces are continually moist. Diurnal temperature changes will also affect water flow in soil. As the air cools at night and the soil surface cools water flow to the surface will be reduced.

Experiments were undertaken initially to determine whether loss of organochlorine insecticides by volatilization were significant and the effect of agricultural management on those losses. Losses may be measured by sampling soils at intervals after application or by air sampling to determine the amount that leaves the field. The measurement of the amount remaining in soil is technically simpler. Loss from soil is caused by volatilization or by biological–chemical degradation. If the biological–chemical degradation rate is known, the difference between this quantity and the total lost is the volatility loss. In practice the method is complicated by the variability of the degradation rate within a relatively small area. For example, a field study of the degradation of carbaryl applied to a watershed showed that the rate of degradation varied with the sampling site and the variability can probably be ascribed to the time required to establish a microbial population at each site [12]. However, the technique has value as a comparative method and could be refined to overcome some inherent difficulties.

The process of volatilization depends on the vapour pressure of the compound and, as was previously discussed, it depends on the nature of the surface on which the residues occur. The vapour pressure of most pesticides is quite low (see Table I) and the measurement of such low vapour pressures has presented difficulties in the past. The methods of determination and the experimental difficulties have been discussed [13]. Two methods which are most satisfactory are the gas saturation method [14] and the gas chromatographic method [15]. The former is based on the measurement of the amount of the substance present in a given volume of air and the latter is based on the determination of gas chromatographic retention times. There are many values reported in the literature for individual compounds and the examples cited here are from a recent critical review [5].

The amount of pesticide in the air leaving a defined area is termed the flux. Evaporation from the surface and movement into the air above the field involve a change of state followed by movement across a layer of air. Movement across this layer depends on molecular properties and it is assumed that molecular diffusion occurs across a region of laminar flow. Subsequently, the vapour moves into a turbulent zone in which molecular properties no longer appear to dominate movement and movement is then described in terms of the eddy diffusion coefficient.

Pesticide flux rate is given by $P = (K_z) (dp/dz)$, where dp/dz is the pesticide vapour density gradient and K_z is the vertical diffusivity coefficient at height z. The vertical diffusivity coefficient, K_z, can be obtained by measurement of the water vapour flux from soil, E. The water vapour flux is obtained by measurement of loss of water from a weighing lysimeter in the field and, since $E = (K_z) (de/dz)$ where (de/dz) is the rate of decrease of water vapour pressure with height, a value for K_z can be obtained. The value of de/dz is obtained by measurement of moisture levels at two or more heights and dp/dz is obtained by measuring the concentration of pesticide in air at several heights. The basic assumption is that pesticide flux may be treated in the same way as water vapour flux. Alternative approaches to the derivation of a value for the vertical diffusivity coefficient involve similar equations that describe the vertical flow of heat or momentum exchange and in each case the value of K_z is assumed to be the same as that used for water vapour.

A report in 1984 [16] is worthy of detailed description because it incorporates much of the experience gained during the period of study.

In 1975 an area 90 m by 180 m (1.6 ha) at Beltsville was used. Woods 200 m to the east and a highway 600 m to the north were the nearest obstructions to wind flow. A mixture of heptachlor, trifluralin and dacthal was applied by means of a tractor mounted sprayer. Within an hour of application air samples were obtained and sampling continued at intervals using a sampling technique described by Turner and Glotfelty [17]. Plugs of porous polyurethane contained in glass filter tubes, protected against light, were attached to the mast of a sampling assembly at 12 heights above in the field. The assembly also contained a Plexiglas surge tank and was connected to a modified shop vacuum cleaner. Flexible tubing connected the sample tubes, surge tank and vacuum generator. The pump was capable of drawing more than 10 m^3/h through the 12 probes, but the flow rate could be controlled at the surge tank. Pesticides were recovered by hexane or benzene Soxhlet extraction and analysed by electron capture gas chromatography. Tests showed that a single plug could trap 98% of the vapour and retain this even when pesticide saturated air continued to be drawn through the plug for 18 h.

Volatilization losses were estimated from the decrease in soil residues and from aerial pesticide flux measurements. The latter were derived by the aerodynamic method which requires the measurement of vapour density gradients, wind velocity, wind direction and air temperature throughout the sampling period. Initial rates of volatilization from moist soil were rapid and were proportional to the saturated vapour densities of the pure compounds. Half the amount of each chemical except dacthal was lost in less than 3 days. Losses from dry soil were much slower until the surface was moistened by dew in the evening. Soil incorporation was very effective in reducing volatilization. For example, half of the applied herbicide, trifluralin (vapour pressure 0.006 Pa), disappeared from a moist soil surface within six hours. Incorporation to a depth of 7.5 cm reduced the loss to 3.4% over a period of 90 days.

Using these techniques, post-application volatility losses of a variety of pesticides were studied. A wide range of rates was found depending on the climatic conditions, the nature of the environmental surfaces, the method of application and the properties of the compound. Losses of up to 90% of the more volatile compounds were observed in 2 to 3 days under warm, humid conditions.

TABLE II. PESTICIDES TRANSPORTED BY AIRBORNE DUST [18]

Compound	Amount found (ppm)
DDT	0.6
Chlordane	0.5
DDE	0.2
2,4,5-T	0.04
Arsenic	26

4.4. Movement in air and redeposition

The Chernobyl accident provided a reminder of the importance of atmospheric movement in the transport and global distribution of pollutants. Sensitive methods of detection make it possible to track the environmental movement of man made organic chemicals, including many pesticides. As is the case with radioactive compounds the detection of pesticides and the extent to which they move are dependent on their stability. Because organochlorine insecticides are resistant to chemical and biological degradation they are frequently detected in environmental monitoring programmes.

Pesticides may be lost from terrestrial surfaces as vapour or in association with particulate matter. Removal from the atmospheric circulation may occur by reaction or deposition. In the atmosphere chemical change may be brought about by photochemical reactions or by reaction with activated molecular species. These may be homogeneous reactions in the vapour phase or heterogeneous reactions of materials adsorbed on particles.

The return of pesticides and other organics to the Earth's surface may occur in a number of ways. They may return as vapour or dry particles. Dissolved material or particles may be returned in rainfall, fog, ice or snow. Measurements of pesticides in rainfall have been made at great distances from the site of application. For example, pesticides in dust from the southern high plains of Texas were transported to Cincinnati, Ohio, where the dust was washed out. Pesticide content is shown in Table II [18].

DDT loss by volatilization and its significance as a pollutant of the oceans was the subject of a National Academy of Sciences report which concluded that as much as 25% of the DDT produced up to that time might have been transferred to the sea and that vapour loss, aerial drift and volatilization of applied material contributed significantly to this [19].

Although pesticides and many other synthetic organic chemicals have relatively low vapour pressures, little consideration was given at the time of their introduction to the possibility that they would enter the atmosphere in quantity as vapour or adsorbed on particulate matter. This mechanism for widespread distribution of residues has exposed many organisms to extremely low pesticide levels and their potential effects on biota have been the subject of prolonged investigations and increased controversy over the continued use of persistent pesticides.

In considering global movement we are primarily concerned with mechanisms by which pollutants may return to Earth. Airborne materials may return by several routes. They may be deposited on terrestrial surfaces as vapour, as dust or adsorbed on particulate matter. They may also be washed out in rainfall, snow or ice or they may be deposited as fog.

An understanding of the atmospheric transport of pollutants requires knowledge of their distribution among the vapour phase, particulates and the aqueous phase. Foreman and Bidleman have examined partitioning of pesticides and pollutants between vapour and particles in ambient air and have devised techniques for its measurement [20].

Rainfall is an important mechanism for redeposition and Glotfelty has reported values for pesticides in rainfall in Maryland. His report indicates that rainfall may be a significant route by which pesticides enter bodies of water [21].

Glotfelty et al. have also studied deposition of pesticides in fogs [22]. When condensation of water vapour occurs fogs are formed containing droplets of 2 to 60 μm. The ratio of droplet volume to the total volume of a cloud is low and the large proportion remains as vapour. Since fogs and clouds are acidic, there is great interest in the origins of this acidity in relation to chemical transformations in the atmosphere.

Recent reports have described the identification of carbonyls, acids surfactants and other compounds in fog [23, 24]. However, the problem of sampling presents major difficulties. A sampler was designed to overcome the problem of obtaining a large volume of fog liquor for analysis capable of sampling both vapour and liquid

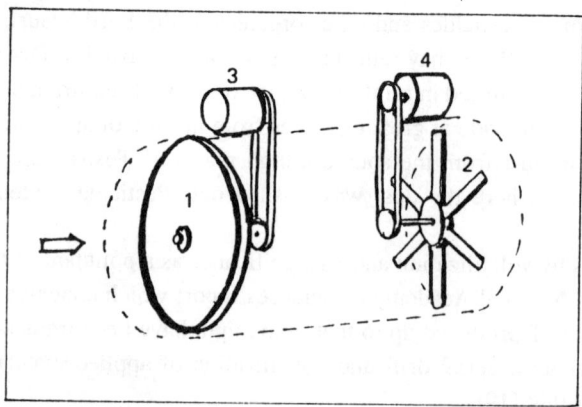

FIG. 2. *High volume atmospheric fog extractor: (1) rotating impaction screen with peripheral channel collector; (2) fan; (3) (4) motors.*

phases of fog. The equipment was built at Beltsville by Drs Glotfelty and Liljedahl in co-operation with Professor Seiber at the University of California, Davis.

To extract the liquor from the fog requires a very high volume sampler because the liquor represents only a very small fraction of the fog. The collector handled 4000 m^3/h of air and it depended on the impaction of fog droplets on a rotating stainless steel screen. Its efficiency was more than 99% for drops larger than 4 µm. During operation 0.2 to 2.0 L of liquor were collected. The design has now been modified to replace the rotating stainless steel screen by a stationary teflon mesh screen fixed at an angle and allowing fog liquor to drain into a trap after droplets impacted the screen (Fig. 2).

A dichotomous sampler was used to sample the interstitial vapour [25]. The total system flow rate was about 0.5 m^3/min and orifices were modified to increase the cut-off diameter to about 8 µm. Organic vapours and particles were trapped and fog droplets were carried out of the system. Particles were filtered out of the droplet free air, which then passed through absorbent traps containing polyurethane and Chromosorb 102.

Dichloromethane extracts were analysed after subfractionation by high pressure liquid chromatography on silica gel. Pesticides were identified by gas chromatography using element selective detectors and mass spectrometry.

The sampler was mounted on a van and samples were collected at a number of locations in California and at Beltsville. The latter is a suburban residential area and the locations in California are primarily agricultural, including cotton, citrus, fruit and grape growing areas and a dairy. It was found that fog liquor samples collected at Beltsville and in California contain pesticides, flame retardants, pesticide alteration products, plasticizers, industrial chemicals and combustion products [22].

TABLE III. PESTICIDES IN FOG [22]

Pesticide	Location	Concentration (ng/L)	
		Fog water	Air ($\times 10^{-3}$)
Diazinon	P	16 000	2.2
	L	22 000	5.3
	B	140	8.6
Parathion	P	12 400	3.2
	C	5 000	0.88
	L	51 400	7.9
Malathion	P	70	<0.03
	C	110	0.02
	L	350	
	B	2 740	1.9
Methyl parathion	B	1 210	7.3
Paraoxon	P	9 000	0.21
	C	950	0.066
	L	184 000	0.25
Diazinon oxon	P	190	<0.03
Atrazine	P	270	<0.2
	C	320	<0.16
	L	700	—
	B	820	3.4
Simazine	P	390	<0.1
	C	110	<0.06
	L	1 200	—
	B	45	<0.04
Pendimethalin	P	1 370	0.64
	C	3 620	3.6
Alachlor	B	1 450	<0.4
Metolachlor	B	1 960	<0.44

Sampling locations: P = Parlier, CA; C = Corcoran, CA; L = Lodi, CA; B = Beltsville, MD.

Little of the pesticide was associated with particulate matter. Beltsville samples were more acidic (pH = 2.42) than those from California. The higher concentrations and greater diversity of pesticides in the California samples reflect more extensive use of pesticides and greater diversity than at Beltsville. Organophosphorus compounds predominate and the oxon analogues of several organophosphorus pesticides were found. Of these paraoxon was most abundant. They are known to be formed by the gas phase reaction of the parent compound with ozone or the reaction may occur as a heterogeneous reaction on dry surfaces.

In addition to organophosphorus compounds, herbicides of several classes were found: s-triazines (simazine, atrazine), dinitroaniline (pendimethalin), and dichloroacetanilides (alachlor, metolachlor). Some data are given in Table III [22].

A more detailed study of the distribution of organic compounds between air and fog water is continuing at Beltsville. The distribution betweeen the aqueous phase and the vapour phase of the fog was calculated in each case and the value compared with that derived from Henry's Law. This states that the ratio of the vapour pressure to the concentration of a pure solute of low solubility is constant, or P = HC_w, where H is the Henry's Law constant, P is the vapour pressure of the solute and C_w is its concentration in the aqueous phase. The distribution coefficients calculated from field measurements did not agree with those derived from literature values [22].

Some pesticides were present in much higher concentrations in the aqueous phase and the investigators termed the relationship between the experimentally measured and the calculated value, the 'enrichment factor'. In some cases, such as that of the herbicides, pendimethalin, it was very large.

To explain these discrepancies it was suggested that interactions among fog components may be involved or that surface active materials may be responsible for the enrichment. The discrepancies also called for a further examination of the physical data used in the calculations and Fendinger and Glotfelty are currently measuring Henry's Law constants for a number of pesticides. Determinations for lindane, alachlor and diazinon have been made using a specially constructed apparatus that employs the wetted wall column technique and is suitable for pesticides with vapour pressures in the 10^{-5} to 10^{-7} mm range [26]. This study and the studies of fog are continuing. They have far reaching implications and emphasize the need to re-examine fundamental physical data. They also reveal that the routes by which pesticides and pollutants may re-enter man's environment are not well understood.

5. CONCLUSIONS

The routes by which many synthetic organic compounds may be transported from their site of application have been described. Pesticides have provided excellent examples and models for study. If the physical and chemical properties of a compound are precisely defined before proposed use in the environment is permitted the potential for adverse environmental effects may be minimized. Although we now have much information concerning the fate and distribution of man made chemicals, there are awesome gaps in our knowledge and there remains the continuing difficulty of satisfactorily combining environmental information when numerical values may differ greatly in orders of magnitude.

CHEMICALS REFERRED TO IN TEXT

Common name	Chemical name
Alachlor	2-chloro-2',6'-diethyl-N-methoxymethylacetanilide
Aldicarb	2-methyl-2-(methylthio)propionaldehyde 0-methylcarbamoyloxime
Atrazine	6-chloro-N^2-ethyl-N^4-isopropyl-1,3,5-triazine-2,4-diamine
Chlordane	1,2,4,5,6,7,8,8-octachloro-2,3,3a,4,7,7a-hexahydro-4,7-methanoindene
Dacthal (chlorthal-dimethyl)	dimethyl tetrachloroterephthalate
DDE	1,1-dichloro-2,2-bis(4-chlorophenyl)ethylene
DDT	1,1,1-trichloro-2,2-bis-(4-chlorophenyl)ethane
Diazinon	0,0-diethyl 0-2-isopropyl-6-pyrimidinyl-4-yl phosphorothioate
Dieldrin	(1R,4S,4aS,5R,6R,7S,8S,8aR)-1,2,3,4,10,10-hexachloro-1,4,4a,5,6,7,8,8a-octahydro-6,7-epoxy-1,4:5,8-dimethanonaphthalene
Endosulfan	(1,4,5,6,7,7-hexachloro-8,9,10-trinorborn-5-en-2,3-ylenebismethylene)sulphite
Heptachlor	1,4,5,6,7,8,8-heptachloro-3a,4,7,7a-tetrahydro-4,7-methanoindene
Lindane (gamma-HCH)	1,2,3,4,5,6-hexachlorocyclohexane (mixed isomers)
Malathion	diethyl (dimethoxyphosphinothioylthio)succinate
Methyl parathion	0,0-dimethyl 0-(4-nitrophenyl)phosphorothioate

Metolachlor	2-chloro-6'-N-(2-methoxy-1-methyethyl)acet-o-toluidide
Parathion	0,0-diethyl 0-(4-nitrophenyl)phosphorothioate
Pendimethalin	N-(1-ethylpropyl)-2,6-dinitro-3,4-xylidene
Simazine	6-chloro-N^2, N^4-diethyl-1,3,5-triazine-2,4-diamine
2,4,5-T	(2,4,5-trichlorophenoxy)acetic acid
Trifluralin	a,a,a-trifluoro-2,6-dinitro-N,N-dipropyl-p-toluidine

REFERENCES

[1] FAO/IAEA Advisory Group, The Application of Nuclear Techniques to Study Environmental Processes Influencing Pesticide Behaviour, IAEA, Vienna, 1984 (Internal report).
[2] PIMENTEL, D. (Ed.), World Food, Pest Losses and the Environment, Westview Press, Boulder, CO (1978).
[3] MELLOR, J.W., ADAMS, R.H., Jr., Chem. Eng. News (23 Apr. 1984) 32.
[4] Farm Chem. (Sept. 1985) 26.
[5] SUNTIO, L.R., SHIU, W.Y., MACKAY, D., SEIBER, J.N., GLOTFELTY, D.E., A critical review of Henry's Law Constants for pesticides, Rev. Environ. Chem. Toxicol. (in press).
[6] GLOTFELTY, D.E., TAYLOR, A.W., ISENSEE, A.R., JERSEY, J., GLENN, S., Atrazine and simazine movement to Wye Estuary, J. Environ. Qual. 13 1 (1984) 115.
[7] WAUCHOPE, R.D., The pesticide content of surface water draining from agricultural fields — A review, J. Environ. Qual. 7 (1978) 459.
[8] Pest. Tox. Chem. News (25 Mar. 1987).
[9] VINTON, A.J.A., YARON, B., NYE, P.H., Vertical transport of pesticides into soil when adsorbed on suspended particles, J. Agric. Food Chem. 31 3 (1983) 662.
[10] TAYLOR, A.W., Post-application volatilization of pesticides under field conditions, J. Air Pollut. Control Assoc. 28 9 (1978) 922.
[11] SPENCER, W.F., CLIATH, M.M., Pesticide volatilization as related to water loss from soil, J. Environ. Qual. 2 (1974) 284.
[12] CARO, J.H., FREEMAN, H.P., TURNER, B.C., Persistence in soil and losses in runoff of soil-incorporated carbaryl, J. Agric. Food Chem. 22 5 (1974) 860.
[13] PLIMMER, J.R., "Volatility", Herbicides: Chemistry, Degradation and Mode of Action (KEARNEY, P.C., KAUFMAN, D.D., Eds), Vol. 2, Marcel Dekker, New York (1976) 892.
[14] THOMPSON, G.W., Techniques of Organic Chemistry, 3rd edn, Vol. 1, Part 1 (WEISSBERGER, A., Ed.), Wiley, New York (1963) Chapter 9.
[15] BIDLEMAN, T.F., Estimation of vapor pressures for nonpolar organic compounds by capillary gas chromatography, Anal. Chem. 56 (1984) 2490.

[16] GLOTFELTY, D.E., TAYLOR, A.W., TURNER, B.C., ZOLLER, W.H., Volatilization of pesticides from fallow soil, J. Agric. Food Chem. **32** 3 (1984) 638.
[17] TURNER, B.C., GLOTFELTY, D.E., Field air sampling of pesticide vapors with polyurethane foam, Anal. Chem. **49** 1 (1977) 7.
[18] COHEN, J.M., PINKERTON, C., "Organic pesticides in the environment", Advances in Chemistry (GOULD, R.F., Ed.), American Chemical Society Publications, Washington, DC (1966) 163.
[19] NATIONAL ACADEMY OF SCIENCES, Chlorinated Hydrocarbons in the Marine Environment, Panel Report, Committee on Oceanography, NAS, Washington, DC (1971).
[20] FOREMAN, W.T., BIDLEMAN, T.F., An experimental method for investigating vapor-particle partitioning of trace organic pollutants, Environ. Sci. Technol. **21** 10 (1983) 2093.
[21] GLOTFELTY, D.E., "Pathways of pesticide dispersion in the environment", Agricultural Chemicals of the Future (HILTON, J., Ed.), BARC Symposium, Vol. 8, Rowman and Allenheld, Totowa, NJ (1985) 425.
[22] GLOTFELTY, D.E., SEIBER, J.N., LILJEDAHL, L.A., Pesticides in fog, Nature (London) **325** (1987) 602.
[23] GROSJEAN, D., WRIGHT, B., Carbonyls in urban fog, ice fog, cloudwater, and rainwater, Atmos. Environ. **17** 10 (1983) 2093.
[24] KAWAMURA, K., KAPLAN, I.R., Capillary gas chromatography determination of volatile organic acids in rain and fog samples, Anal. Chem. **56** 9 (1984) 616.
[25] SOLOMON, P.A., MOYERS, J.L., FLETCHER, R.A., High volume dichotomous virtual impactor for the fractionation and collection of particles according to aerodynamic size, Aerosol Sci. Technol. **2** (1983) 455.
[26] FENDINGER, N.J., GLOTFELTY, D.E., A laboratory method for the determination of air–water Henry's Law Constants for several pesticides, Agrochem. Div., Amer. Chem. Soc., Natl Meeting, New Orleans, LA (1987) (abstract only).

EMPLEO DE PLAGUICIDAS Y EXPERIENCIAS CON TECNICAS RADISOTOPICAS EN UN PAIS EN VIAS DE DESARROLLO*

J. ESPINOSA-GONZALEZ, F. RAMON,
E. BORRERO de SAIZ, M. NAVARRO,
D. ROVIRA, A. GUERRA
Instituto de Investigación Agropecuaria
  de Panamá (IDIAP),
Universidad de Panamá,
El Dorado, Panamá, Panamá

Abstract–Resumen

USE OF PESTICIDES AND EXPERIENCE OF APPLYING RADIOISOTOPE TECHNIQUES IN A DEVELOPING COUNTRY.
   An evaluation is made of the use of pesticides by Panamanian farmers in a tropical environment, also covering pesticide residues in plant and animal products, man and soil. In addition, experience with radioisotope techniques is described. Chemical control is common practice among farmers. Each year, 5000 to 6000 t of pesticides are used, especially in horticulture and banana cultivation. Herbicides and insecticides predominate in terms of quantity, and fungicides in terms of frequency of application. Use of the so-called persistent organochlorines over the last few decades has led to the presence of residues in plant and animal products in amounts less than 2.2 and 0.1 mg/kg for DDT and lindane, respectively. An average of 11 mg of DDT per kilogram of fat has been detected in the population; about 50% of the persons handling agrochemicals showed direct exposure. Taking into account local practices and tropical conditions, an evaluation is being made of widely used pesticides (maneb, paraquat and 2,4-D) labelled with ^{14}C. The studies have yielded additional information on the behaviour and the residues of these important additives in the environment and in fruits.

EMPLEO DE PLAGUICIDAS Y EXPERIENCIAS CON TECNICAS RADISOTOPICAS EN UN PAIS EN VIAS DE DESARROLLO.
   Se presenta un diagnóstico sobre el uso de plaguicidas por agricultores panameños en el medio tropical, incluyendo residuos de estas sustancias en los productos agropecuarios, el hombre y el suelo. Además, se informa sobre experiencias con técnicas radisotópicas. El control químico es una práctica común entre los agricultores. Anualmente, se emplean de 5000 a 6000 toneladas de plaguicidas, especialmente en los ecosistemas hortícolas y de musáceas. Los herbicidas y los insecticidas han sido los de mayor uso en cuanto a cantidad, pero los fungicidas predominan en cuanto a frecuencia de uso. El empleo de los denominados persistentes

---

   * Investigación llevada a cabo con el apoyo del OIEA bajo el Contrato de Investigación N° 4015/RB.

clorados durante las últimas décadas ha traido la presencia de residuos en los frutos agropecuarios en cantidades menores a los 2,2 y 0,1 mg/kg de DDT y lindano respectivamente. En la población se detectó un promedio de 11 mg de DDT por kg de grasa; las personas que manipulan los agroquímicos mostraron quedar expuestas de forma directa en aproximadamente un 50%. Considerando las prácticas locales y las condiciones tropicales, se están evaluando plaguicidas de amplio uso (maneb, paraquat y 2,4-D) marcados con C 14. Los estudios han permitido ampliar conocimientos sobre el comportamiento y los residuos de estos importantes insumos en el ambiente y en frutos.

1. INTRODUCCION

Nuestro país, al igual que otras naciones en vías de desarrollo, viene cursando durante esta época por una serie de problemas no solo de índole económica, social y política, sino también de tipo ecológico; en ello inciden vectores de magnitud mundial en forma determinante. La débil participación en la generación tecnológica de nuestros países no acelera el desarrollo deseado y abre una brecha cada vez mayor en relación a los países industrializados.

Nuestra economía, basada en el servicio y el intercambio comercial, no llena usualmente las necesidades básicas de la creciente población, por lo que los gobiernos han realizado grandes esfuerzos tratando de fortalecer el sector alimentario a través de una agricultura tecnificada que ofrezca mejores rendimientos y una mayor productividad. Como fruto de esta filosofía se han introducido una serie de tecnologías y materias procedentes de países más avanzados de forma directa, sin la validación correspondiente y olvidando los factores propios, a veces específicos del ambiente tropical y su comunidad biótica.

La participación de sustancias químicas en las actividades agropecuarias de nuestro medio han mejorado la productividad, sin lugar a dudas, pero se emiten críticas por parte de personas que ven en el uso de los agroquímicos una cierta desestabilización del equilibrio ecológico. Por otro lado, los agricultores se quejan de la calidad y los elevados precios de esos productos químicos.

Si bien los agroquímicos se están empleando en nuestro ambiente agrícola en cantidades considerables, es magra la información existente sobre su comportamiento degradativo y la movilidad en nuestros agroecosistemas. Teóricamente, cabe esperar que las drásticas condiciones del trópico, es decir una elevada temperatura, una elevada radiación solar con gran contenido de rayos ultravioletas y una gran diversificada acitividad microbiana, así como las condiciones edafológicas, entre otros, causarán una mayor velocidad degradativa que los ecosistemas templados.

2. EL USO DE LOS PLAGUICIDAS

Los agricultores de nuestro país, que cultivan predominantemente granos básicos (arroz, maíz), musáceas (banano, plátano), hortalizas, caña de azucar y pastos, vienen haciendo uso de los plaguicidas desde hace varios decenios. Sin embargo, durante los últimos tres lustros es cuando se difunde el empleo de estas sustancias en el medio agrícola (Fig. 1). Actualmente, la mayoría de los productores los emplean en sus actividades agrícolas con la intención y la confianza de prevenir o controlar las principales plagas y enfermedades o las malezas (Cuadro I). Las 5000 a 6000 t de plaguicidas usadas en nuestra agricultura proceden generalmente de los EE UU y de Europa, pues la producción nacional es más bien de baja cuantía (10%) [1].

En el proceso productivo, los pequeños y medianos agricultores invierten usualmente alrededor de un 25% en agroquímicos sobre el costo total; de este porcentaje entre el 20 y el 50% es dedicado al control químico, según el cultivo y la época. Evidentemente, los agricultores o empresas con equipo altamente tecnificado invierten mayores sumas.

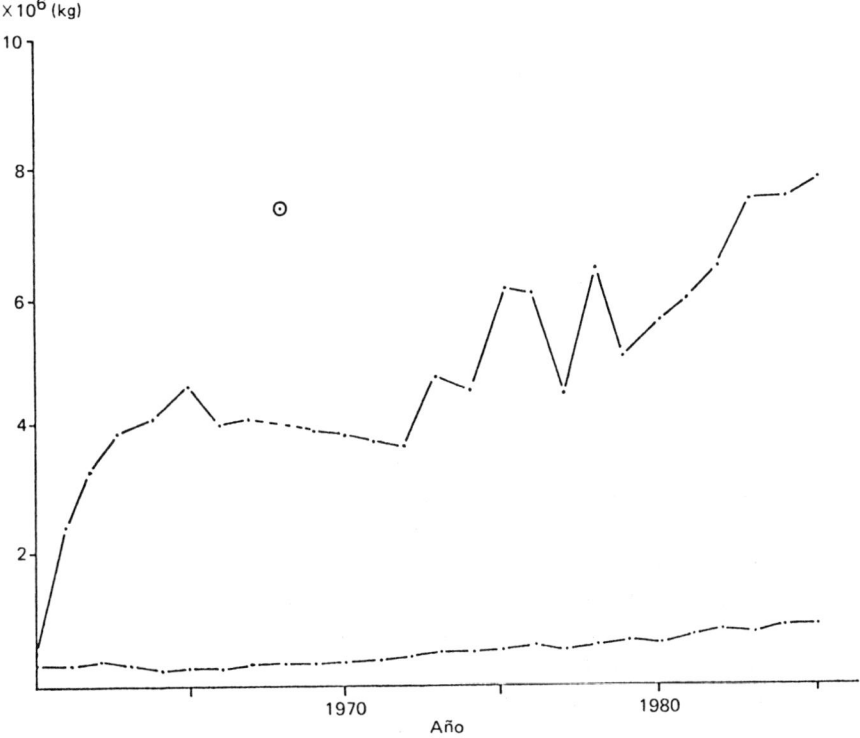

FIG. 1. Demanda de plaguicidas en Panamá.

CUADRO I. PLAGUICIDAS EMPLEADOS EN PANAMA[a]

Plaguicida	Cantidades utilizadas			
	Período		Período	
	(kg)	(%)	(kg)	(%)
Insecticidas	*4 994 614*	*25*	*1 034 619*	*15,9*
(Nematicidas)	2 050 927	41,1	535 131	51,7
Carbamatos	2 080 303	41,7	46 829	4,5
Fosfatos	1 500 225	30	562 779	54,4
Organoclorados	972 997	19,5	?	
Piretroides	?		30 343	2,9
Otros (no especificados)	439 526	8,8	354 926	34,3
Herbicidas	*11 782 292*	*59*	*2 252 639*	*34,7*
Anilidas	7 035 270	59,7	455 610	20,2
Fenóxicos	2 786 971	23,7	545 830	24,3
Piridilos	304 743	2,6	174 484	7,8
Ureas	1 151 538	9,8	92 809	4,1
Otros (no especificados)	801 743	4,2	1 158 390	43,2
Fungicidas	*1 796 437*	*9*	*1 810 407*	*27,9*
No especificados	*1 412 557*	*7*	*1 401 922*	*21.5*
Total	*19 985 900*	*100*	*6 499 587*	*100*

[a] Datos estadísticos de La Contraloría General de La República, Comercio Exterior (1970–1986).

La dosificación en las aplicaciones, predominantemente preventivas, son variables y dependientes de diversos factores que involucran aspectos económicos del agricultor. Los precios de los plaguicidas son altos en relación a los sitios agrícolas de los principales países fabricantes; en ello inciden el flete, los costos y las políticas de comercialización. Evidentemente, los agricultores de los países industrializados reciben una serie de ventajas, como la garantía de calidad y precios razonables y estables, por la cercanía a los centros de elaboración. Además, el clima templado ayuda a la preservación de las materias activas inestables.

Los ecosistemas hortícolas y de musáceas son rubros que están caracterizados por una adición frecuente y en cantidades considerables de químicos. El 60% y más de la cantidad asperjada se precipita sobre el suelo (baja eficiencia en el método de aplicación). Alteraciones del agroecosistema han sido causadas por fungicidas cúpricos que fueron adicionados durante un largo período en centros específicos, cuyo efecto es una toxicidad en granos básicos superior a 500 mg/kg de suelo. Adicionalmente, durante la introducción de los altamente tóxicos organofosfatos, se presentaron numerosas toxicosis en obreros agricultores, debido al bajo conocimiento técnico necesario en el buen y correcto manejo de estas sustancias. Las nebulizaciones o aspersiones se difunden fácilmente en la atmósfera tropical, que es muy fluida; de allí que la contaminación sea un fenómeno frecuente. Esta realidad la verificamos con gran facilidad en aquellas personas hipersensibles a ciertas sustancias.

Nuestros agricultores tienen a su disposición un centenar y medio de fórmulas de cualquier y todo tipo: organoclorados, organofosfatos, carbamatos, piretroides, atrazinas, hormonales, etc. Los herbicidas son los de mayor uso en cuanto a cantidad, siguiendo en otro orden los insecticidas y fungicidas. El número de aplicaciones es variable, pero asciende en cultivos de manejo intensivo (hortalizas) hasta 13 veces/ciclo cuando se pretende controlar insectos y enfermedades micóticas; de las dosis aplicadas, que oscilan entre 0,2–1,0 kg/ha (insecticidas) y 0,3–4 kg/ha (fungicidas), se estima que anualmente se depositan en ciertas áreas hasta 130 kg de tales sustancias por ha. En numerosos casos, los agricultores aplican productos químicos mezclados con el objeto de ahorrar mano de obra, costo y tiempo, pero en algunos casos se presenta fitotoxicidad.

También en las zonas altas, donde la temperatura es benigna, el manejo de los agrotóxicos se efectúa sin las precauciones y el procedimiento debidos para el usuario y el medio ambiente. Es poco común ver a un roceador agrícola equipado con indumentaria de protección. Evidentemente, esta realidad lleva de forma intrínseca a la exposición aguda, fenómeno que hemos constatado a través de la determinación de colinesterasa antes y posterior a la aspersión de esteres organofosforados. En más del 50% de los roceadores evaluados, los niveles de colinesterasa se redujeron sobre el 15% del valor normal después de un período de aplicación de 2–3 horas (valor promedio: 2.512 U/mL antes de labor — valor postexposición promedio: 2.298 U/mL).

3. RESIDUOS

Un manejo correcto de los tóxicos en el medio agrícola no debería traer efectos negativos de considerable magnitud. Sin embargo, llevar a cabo este proceso requiere de un usuario bien capacitado, muy consciente y de ciertos aspectos que son realizables, pero que por razones antropogénicas no se efectuan normalmente. En cierto grado, el manejo correcto, adecuado y oportuno de los plaguicidas es sofisticado. Los agricultores locales han venido empleando en considerable cuantía

CUADRO II. RESIDUOS DE INSECTICIDAS EN PRODUCTOS AGROPECUARIOS (en mg/kg)

Producto	HCB	Alfa-HCH	Lindano	Clordano	DDT	Dieldrina	Heptacloro
Bovino	0,039	nd	0,072	0,16	2,1	0,022	nd
Porcino	0,020	0,013	0,021	nd	0,25	0,22	0,261
Aviar	nd	nd	0,028	nd	0,11	nd	0,010
Queso fresco	nd	nd	0,028	nd	0,019	nd	nd
Arroz	nd	nd	nd	nd	nd	nd	nd
Cebolla	nd	nd	nd	nd	nd	nd	nd
Tomate	nd	nd	nd	nd	nd	nd	nd
Papa	nd	nd	nd	nd	nd	nd	nd
Sensibilidad	0,001	0,001	0,001	0,004	0,010	0,005	0,002

nd: no detectado.

CUADRO III. RESIDUOS DE DDT EN LA POBLACION PANAMEÑA Y SU DISTRIBUCION PORCENTUAL

DDT (mg/kg grasa)	Frecuencia (%)
<1	11
1–4,9	31
5–9,9	23
10–20	18
>20	17

estas sustancias, como se indicó anteriormente, desde buen tiempo atrás. Actualmente, el uso de los insecticidas llamados persistentes clorados ha disminuido, empero éstos se aplicaron durante los años 1950-1970 tanto en el control de vectores de enfermedades tropicales como en la agricultura. Como es bien conocido, estas sustancias han sido y siguen siendo criticadas fuertemente por ciertos grupos de personas.

Con el objeto de conocer mejor la situación de los plaguicidas y la presencia de residuos en el ambiente tropical, se han efectuado estudios en productos agropecuarios y lípidos de la población con técnicas analíticas modernas. El DDT y lindano son las sustancias que se presentaron con mayor frecuencia (30%) en las muestras estudiadas, en cantidades inferiores a 2,2 y 0,1 mg/kg respectivamente. Estos valores están dentro de los límites permisibles establecidos por FAO/OMS [2]. Además, se percibió un efecto de bioacumulación en determinado eslabón de la cadena alimentaria del hombre. Ello es válido sobre todo para el ganado (bovino, porcino y aviar), por lo que se deduce que se trata de un fenómeno de interacción hombre, medio ambiente y plaguicidas (Cuadro II).

La exposición de los roceadores agrícolas a organoclorados por ingesta directa y/o de alimentos fue evaluada en base al DDT y sus metabolitos en los lípidos de la población femenina (leche) y masculina (grasa). Los resultados indican que el 65% de la población femenina estudiada contenía menos de 10 mg mientras que el 17% mostró entre 21 y 90 mg de DDT total por kg de grasa. El contenido promedio en las personas estudiadas fue de 11 mg/kg (Cuadro III) [3].

4. EXPERIENCIAS CON TECNICAS RADISOTOPICAS

Con el objeto de determinar el destino, los residuos, y para conocer el comportamiento de plaguicidas de amplio uso por la mayoría de agricultores (pequeños y medianos), considerando sus prácticas de manejo y las condiciones ecológicas donde se cultivan importantes rubros agrícolas, se están realizando ensayos con sustancias activas marcadas con C 14. A continuación se presenta una descripción de los estudios y algunos resultados preliminares.

4.1. Maneb C 14 en tomate

Se asperjaron plantas de tomate (*Lycopersicum esculatum*) con maneb C 14 (actividad específica = 117,72 μCi/mg) mezclado con maneb 80WP comercial, siguiendo la práctica local de los pequeños agricultores.[1] El tratamiento consistió en la aspersión de una dosis correspondiente a 0,8 kg/ha cada 10-12 días durante varias semanas en el transcurso de la estación lluviosa (septiembre-noviembre) hasta

---

[1] 1 Ci = 3.70×10^{10} Bq.

CUADRO IV. RESIDUOS DE MANEB C 14 EN TOMATE

	Maneb C 14 total (mg/kg)	ETU (mg/kg)	Maneb C 14 ligado (mg/kg)	Otros (mg/kg)
Fruto	0,57	0,08	0,30	0,19
Hojas	1,28	0,20	0,57	0,51
Tallo	0,36	0,05	0,28	0,03

CUADRO V. RESIDUOS DE LINDANO C 14 EN ARROZ ALMACENADO

Tiempo post-tratamiento (meses)	Superficie (mg/kg)	(%)	Grano (mg/kg)	(%)	Cáscara (mg/kg)	(%)
0 (hasta 24 h)	13,4	14,2	7,6	8,0	73,6	77,8
0,5	11,9	20,2	6,1	10,4	40,9	69,4
1	7,4	12,5	5,4	9,2	29,0	49,2
2	6,5	23,8	4,8	17,6	16,0	58,6
4	6,3	29,5	4,3	20,1	10,7	50,4
6	2,5	20,3	4,0	32,3	5,9	47,5
9	2,1	26,5	2,7	33,5	3,2	40,1

10 días antes de la cosecha. Se colectaron muestras del fruto, hojas y tallo, que fueron luego analizadas por maneb y metabolitos a través de la cromatografía en capa fina, la autoradiografía, la combustión húmeda y el centelleo líquido.

Los resultados indican una distribución de 2-3 veces mayor en el follaje que en el fruto y el tallo, respectivamente. El maneb no persiste como tal, y el metabolito de importancia toxicológica etilenotiourea (ETU) se presentó en el fruto con 0,08 mg/kg, lo que representa el 14% de los residuos. El resto se compone de otros metabolitos sulfurados (32,8%) y de residuos ligados (52,6%) del maneb (Cuadro IV).

4.2. Lindano C 14 en arroz almacenado

El lindano C 14 (actividad específica = 60 μCi/0,33 mg) fue aplicado en formulación para espolvoreo en arroz almacenado según la práctica de pequeños y medianos productores. El objetivo principal del ensayo era conocer aspectos de residualidad de este insecticida en un grano básico de la dieta del panameño.

Los residuos superficiales, separables y ligados, se evaluaron hasta 9 meses después de la aplicación. La cáscara contenía inicialmente un 78%, la mayor cantidad de residuos (74 mg/kg); el 14% mostró estar sobre la superficie, mientras que en el grano se detectó el 8% de la cantidad aplicada. Tras los 9 meses del tratamiento, el grano mostró contener 2,7 mg/kg del residuo de lindano, lo que representa un tercio del residuo total, y la cáscara contenía a esta fecha el 40% (Cuadro V).

4.3. Paraquat C 14 y 2,4-D C 14 en maíz

Los activos de estos herbicidas marcados con C 14 (actividad específica = 425 μCi/mg y 126,8 μCi/mg respectivamente) fueron mezclados con productos comerciales y aplicados en la dosis de 0,4 kg/ha, en parcelas experimentales de maíz, según la práctica de los pequeños productores en una zona tropical de gran actividad agrícola. Muestras de suelo, maíz, hojas y tallo están siendo evaluadas con respecto a la disipación, la degradación y la distribución de residuos.

AGRADECIMIENTOS

Se expresa especial agradecimiento al Organismo Internacional de Energía Atómica por la colaboración brindada.

REFERENCIAS

[1] ESPINOSA-GONZALEZ, J., "Elaboración, comercialización y empleo de plaguicidas en la República de Panamá", Reunión sobre Plaguicidas en Latinoamérica, Argentina, Organización de las Naciones Unidas para el Desarrollo Industrial, Viena (1983) 23.

[2] ESPINOSA-GONZALEZ, J., Informe Detallado del Estudio sobre la Contaminación de Frutos Agropecuarios por Plaguicidas, Instituto de Investigación Agropecuaria de Panamá, Panamá (1985) 19.

[3] ESPINOSA-GONZALEZ, J., WERNER, G., THIEL, R., "Residuos de insecticidas en lípidos de mujeres y hombres panameños", III Congreso Científico Nacional, Universidad de Panamá, Panamá (1986).

EFFECTS OF ^{35}S-DIMEHYPO PESTICIDE ON THE AGRICULTURAL ENVIRONMENT AND ECOSYSTEM

Z.Y. CHEN, C.Y. MI, D.C. YE, B. CHEN
Nanjing Agricultural University,
Nanjing, Jiangsu, China

Abstract

EFFECTS OF ^{35}S-DIMEHYPO PESTICIDE ON THE AGRICULTURAL ENVIRONMENT AND ECOSYSTEM.

Dimehypo is a new type of insecticide of the nereistoxin family of compounds which has been developed and manufactured in China. To appraise its environmental safety, radioisotope tracer techniques were applied to investigate the effects of ^{35}S labelled dimehypo on the agricultural environment and ecosystem. The results revealed its low adsorption and high mobility in soil, stability in soil and water, and slow rate of degradation. The main product of degradation, nereistoxin has a lower mobility than the parent compound in the soil. Sulphur-35-dimehypo was taken up by the grass carp (*Ctenopharyngodon indellus*) along with bait or via the respiratory tracts and was excreted rapidly after the fish were removed from the contaminated water. Liquid and granulated formulations of ^{35}S-dimehypo were fed to quails and fowls, respectively. The radioactivity was excreted rapidly in faeces and urine. The low partition coefficient of the insecticide in caprylalcohol–water suggested no (or low) accumulation in the adipose tissue of the organism. The release of its active ingredients from the granules prepared with porcelain clay or clay soil is prompt and complete.

1. INTRODUCTION

The environmental fate of a pesticide is one of the important aspects to be considered in making a comprehensive assessment [1]. Dimehypo is a new kind of insecticide of the nereistoxin family of compounds which has been developed and manufactured in China [2]. The paper discusses the investigations carried out on the environmental behaviour of ^{35}S labelled dimehypo, i.e. its adsorption, mobility and degradation in soil, stability in water, absorption and excretion by the grass carp, fowl and quail, distribution coefficient in caprylalcohol–water, and release from different granular formulations. Sulphur-35-dimehypo, shown in Fig. 1, was developed by the Atomic Energy Utilization Institute, China Agricultural Science Academy. Thin layer chromatography (TLC) and autoradiography showed that ^{35}S-dimehypo and ^{35}S-nereistoxin were radiochemically pure.

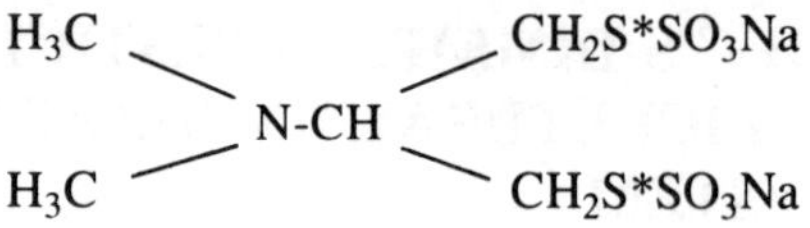

$$H_3C\diagdown\diagup CH_2S^*SO_3Na$$
$$N\text{-}CH$$
$$H_3C\diagup\diagdown CH_2S^*SO_3Na$$

FIG. 1. *Sulphur-35 labelled dimehypo; asterisk indicates ^{35}S-label. Sulphur-35 labelled nereistoxin is derived from the oxidation of ^{35}S-dimehypo.*

2. ADSORPTION OF ^{35}S-DIMEHYPO IN SOIL

2.1. Methods

Sulphur-35-dimehypo aqueous solution with a known specific activity was applied to two different types of soil and water was added to each. When the distribution of radioactivity between the soil and aqueous phases reached equilibrium, the radioactivity in the aqueous phase was measured. According to the change in the concentration of radioactivity, the soil adsorption ratio was estimated [3, 4].

2.2. Results

The soil adsorption ratios were found to be 2.3 and 1.5% in the two soil phases, indicating that the soil samples adsorbed little ^{35}S-dimehypo.

3. MOBILITY OF ^{35}S-DIMEHYPO IN SOIL

3.1. Methods

The methods used for the study of ^{35}S-dimehypo mobility in the soil included soil column chromatography [5, 6], soil TLC and autoradiography [7, 8].

3.2. Results

Analysis of the radioactivity in each zone of the soil column revealed that the ^{35}S-dimehypo located at the bottom of the column moved up with water to the 0 to 2 cm zone located at the top of the column. The radioactivity in this zone accounted for 63.24% of the total radiosulphur and the radioactivity in the 0 to 4 cm zone was 74.38%. The autoradiograph of a cross-section of the soil column showed that ^{35}S-dimehypo was located in the upper zone. This is in agreement with the radioanalytical results shown above.

The autoradiograph of the soil thin layer chromatogram showed that ^{35}S-dimehypo had migrated with water to the solvent front (Rf value was 1). These results proved that ^{35}S-dimehypo moved easily with water.

4. DEGRADATION OF ^{35}S-DIMEHYPO IN THE SOIL

4.1. Methods

The experiment was conducted in the laboratory and imitated a dry or flooded field. Dry soil was adjusted with distilled water to 60% of the field capacity. Flooded soil was prepared as follows: the soil sample was put in a test tube (20 cm × 2.5 cm) and water was then added to form a 3 cm water layer on top of the soil layer.

First, the dry and flooded soils were pre-incubated for 30 days. Then a 2 mL solution containing 1 µCi of ^{35}S-dimehypo was added and mixed with each soil.[1] All samples were incubated at room temperature in the dark. Duplicate aliquots from each sample were analysed at 5, 15, 30, 45 and 60 days. The radioactivity was extracted, the extracts filtered and the filtrates concentrated and analysed by TLC.

4.2. Results

Analysis of the soil samples 30 and 60 days after incubation indicated that no degradation had occurred. The results indicated that ^{35}S-dimehypo was quite stable in the soil.

5. STABILITY OF ^{35}S-DIMEHYPO IN WATER

5.1. Methods

These experiments were designed to study the stability of aqueous solutions of ^{35}S-dimehypo to thermal and microbial degradation. In the first part of the experiment an aqueous solution containing 2.4 mg/mL (1.2 µCi/mL) of ^{35}S-dimehypo was allowed to stand at 8 to 10°C and 35 to 40°C in the dark. In the second part a 0.155 mg/mL (0.8 µCi/mL) solution of ^{35}S-dimehypo was prepared in double distilled water and in a water extract of anaerobic soil. These solutions were also allowed to stand in the dark at room temperature. The water solutions from these experiments were analysed by TLC to determine the rate and nature of degradation.

---

[1] 1 Ci = 3.70×10^{10} Bq.

5.2. Results

The results indicated that ^{35}S-dimehypo was stable in the water and degraded very slowly. No degradation products were detected after 60 days of incubation. However, when incubated in aqueous solution at 35 to 40°C, degradation products were detected after 90 days. On the other hand, no degradation was detected at 8 to 10°C, even when incubated for 125 days. The type of water used in these experiments, i.e. distilled water versus water extracted from anaerobic soil, had no effect on the degradation of ^{35}S-dimehypo, indicating the stability of this compound to anaerobic microorganisms that may have been present in the soil.

6. ABSORPTION AND EXCRETION OF ^{35}S-DIMEHYPO BY THE GRASS CARP

6.1. Materials and methods

Fifteen day old grass carp (*Ctenopharyngodon indellus*), 8 to 11 cm in body length and 5 to 8 g body weight, were used in this experiment. Fourteen fish were transferred to a glass vat containing 10 L of water and air was bubbled into the water to ensure an adequate supply of oxygen.

The experiment comprised two treatments. In the first treatment the fish were allowed to feed on fish bait treated with ^{35}S-dimehypo. The bait was prepared by breeding *Wolffia arrhiza* Wimm. for 3 to 9 days in a water solution containing 77.6 ppm (0.03 μCi/mL) of ^{35}S-dimehypo. The contaminated bait was washed with clean water before introducing it to the fish. After allowing the fish to feed on the labelled bait for 3 to 9 days they were transferred to clean water to study their excretion of radioactivity.

In the second treatment the fish were kept for 5 days in water containing 5.04 ppm (0.002 μCi/mL) of ^{35}S-dimehypo to study their absorption of the insecticide. After 5 days of exposure to the radiolabelled insecticide the fish were transferred to clean water to study their excretion of dimehypo. These treatments are shown in Table I.

6.2. Results

6.2.1. Absorption and accumulation of ^{35}S-dimehypo

On day 5 after the fish were exposed to contaminated water, the radioactivity in different parts of their bodies was determined; the results are given in Table II.

The concentration of ^{35}S-dimehypo derived radioactivity in the fish bodies showed that the grass carp obviously absorbed a considerable amount of ^{35}S-

TABLE I. EXPERIMENTS WITH ^{35}S-DIMEHYPO IN FISH

Treatment	Bait	Water
I	*Wolffia arrhiza* Wimm. (^{35}S-dimehypo labelled)[a]	Clean water
II	Common *Wolffia arrhiza* Wimm.	Water polluted by ^{35}S-dimehypo[b]

[a] *Wolffia arrhiza* Wimm. was incubated in a ^{35}S-dimehypo solution (0.03 µCi/mL, 77.6 ppm) for 3 to 9 days, and washed with clean running water.
[b] The final concentration was 0.002 µCi/mL (5.04 ppm).

dimehypo from the labelled foodstuff or via the respiratory tracts. Table II also shows that the uptake resulting from foodstuff was much larger than that from respiration. In treatment I the concentration of radioactivity in various tissues of the fish bodies was lower than that in the bait, indicating that only some of the ^{35}S-dimehypo in the foodstuff had accumulated in the fish bodies and that most of it had been excreted. The concentration of radioactivity in the vat water increased to 218 counts·min^{-1}·mL^{-1}. Therefore, ^{35}S-dimehypo does not seem to accumulate in the grass carp. The highest levels of ^{35}S-dimehypo were found in the viscera.

6.2.2. Excretion of ^{35}S-dimehypo

The grass carp was fed in the environment contaminated with ^{35}S-dimehypo for 5 days. Having absorbed and accumulated ^{35}S-dimehypo, they were then placed in clean water. The radioactivity concentrations measured in the fish are listed in Table III.

The results showed that the fish absorbed and accumulated ^{35}S-dimehypo. When they were taken from the contaminated water and transferred to the clean water, the water became radioactive. The radioactivity increased gradually with time; meanwhile, in fish tissues it decreased. The amount of ^{35}S-dimehypo in the fish body had decreased by 50% 5 days after they were removed from the contaminated material. The decrease rates differed: meat > bone > viscera. This indicated that the ^{35}S-dimehypo which had accumulated in the fish body was excreted gradually.

7. GRANULATED ^{35}S-DIMEHYPO AND ITS RELEASE PROPERTIES

Dimehypo has been shown to have considerable toxicity to silkworms [9]. To prevent the contamination of mulberry leaves with dimehypo it is preferable to apply

TABLE II. ABSORPTION AND ACCUMULATION OF ^{35}S-DIMEHYPO BY THE GRASS CARP

Treatment[a]	Concentration of ^{35}S-dimehypo related radioactivity in vat water (counts·min^{-1}·mL^{-1})		Concentration of radioactivity in fish tissue					
			Flesh		Viscera		Bone	
	Initial	Final	CR[b]	CC[c]	CR[b]	CC[c]	CR[b]	CC[c]
I	0	218	3785	0.17	7467	0.34	5731	0.26
II	2150	2102	1368	0.65	3389	1.61	868	0.41

[a] Treatment I: The contaminated material was *Wolffia arrhiza* Wimm. (fish and foodstuff) labelled with ^{35}S-dimehypo. Treatment II: the contaminated material was ^{35}S-dimehypo aqueous solution (fish vat water).

[b] CR = concentration of radioactivity = counts·min^{-1}·g^{-1}.

[c] CC = concentration coefficient = $\dfrac{\text{concentration in the fish body}}{\text{concentration in the vat water}}$

For treatment I: CC = $\dfrac{\text{concentration in the fish body}}{\text{concentration in the foodstuff}}$.

TABLE III. FATE OF ^{35}S-DIMEHYPO IN THE GRASS CARP

	Treatment[a]		I				II			
	Excretion time (d)		0	2	5	7	0	2	5	7
Vat water	Concentration (counts·min^{-1}·mL^{-1})		0	25	46	625	0	13	25	25
Flesh	Concentration (counts·min^{-1}·g^{-1})		3785	1986	1664	1555	1368	604	419	265
	Decrease (%)		0	47.5	56.0	58.9	0	55.9	69.4	80.6
Viscera	Concentration (counts·min^{-1}·g^{-1})		7467	5925	3991	3698	3389	1725	1302	1295
	Decrease (%)		0	20.6	46.5	50.5	0	49.1	61.6	61.8
Bone	Concentration (counts·min^{-1}·g^{-1})		5781	5150	2887	2681	868	—	817	111
	Decrease (%)		0	10.9	50.1	53.6	0	—	5.9	37.2

[a] Treatments are the same as those described in Table I.

this insecticide as a granular formulation instead of a solution when treating farmlands in mixed rice–mulberry areas. In this experiment a number of granular formulations of ^{35}S labelled dimehypo were prepared by using different carriers and the adsorption and release of this compound from these carriers were studied.

7.1. Materials and methods

The materials used as granular carriers included pottery clay, porcelain clay, belozem, tender clay, expanded moist clay, clay, kaolin, gangue powder and red brick powder.

For preparing the formulation of ^{35}S-dimehypo, 2 mL of ^{35}S-dimehypo solution (14 210 counts·min^{-1}·mL^{-1}, 284 counts·min^{-1}·µg^{-1}) were mixed with 0.5 g of a carrier for 2 hours. The ^{35}S-dimehypo concentration in the liquid phase was determined. The relative adsorption ratio of the carrier was calculated using the equation

$$R = \frac{(C_i - C_f)\, V/M}{C_g} \times 100\%$$

where R is the relative adsorption ratio, C_i and C_f are the initial and final concentrations in the liquid phase, respectively (mg/mL), C_g is the concentration in the granular formulation (µg/g (wt/wt)), V is the volume of the liquid phase (mL) and M is the carrier mass (g).

Preparations of granulated ^{35}S-dimehypo, with porcelain clay or clay containing 3% ^{35}S-dimehypo, were mixed by hand using an extrusion technique. The granules were 1 to 1.2 mm in diameter and 5 to 8 mm in length.

To study the release of the effective component in the granular agent, granules of different formulations of ^{35}S-dimehypo were embedded in the soil thin layer plates. After 7 days of constant leaching, these plates were dried and autoradiographed under X-ray film. From the autoradiographs, the ^{35}S-dimehypo release properties in the granular agents were studied.

7.2. Results

The adsorption ratios of porcelain clay, pottery clay, expanded moist clay, clay and gangue powder, which can be used as filler in granulated ^{35}S-dimehypo, are quite low.

The soil TLC and autoradiography results showed that the effective component in the granular agents was completely released after 3 to 4 days of constant leaching. Since no radioactive spots coincided with the areas on the soil plates containing the granules, release of ^{35}S-dimehypo from the granules was considered complete.

8. ABSORPTION AND EXCRETION OF ^{35}S-DIMEHYPO IN FOWL AND QUAIL

Dimehypo has a high acute toxicity to quail [10]. Ingestion of sufficient dimehypo granules by fowls and other birds may cause sickness or even death. Thus, in this experiment liquid and granulated formulations of ^{35}S-dimehypo were fed to fowls and quails and their excretion of these compounds was studied.

8.1. Materials and methods

One male and one female quail and fowl were raised for 1 week. The quails were fed a 3.77 mg/mL (0.214 µCi/mL) solution of ^{35}S-dimehypo, with a dose of 15 mg/kg per bird; this is equivalent to 1/2 of the LD_{50}. Each fowl was fed granules containing 4.8% ^{35}S-dimehypo (30.18 µCi/g) on porcelain clay. The faeces and urine of the dosed birds were collected 2, 4, 8, 12, 24, 48, 72, 96 and 144 hours after treatment and the radioactivity was extracted with acidified water. One millilitre of the extract was mixed with the liquid scintillation solution (PPO 0.5%, POPOP 0.05%) in a xylene/Triton X-100 spectrometer. Extracts of the excrement and urine from untreated birds were used to dilute the ^{35}S-dimehypo for quench correction. The excretion ratio was calculated using the equation

$$\text{Excretion ratio} = \frac{\text{radioactivity excreted (counts/min)}}{\text{radioactivity fed (counts/min)}} \times 100\%$$

8.2. Results

The results showed that ^{35}S-dimehypo could be excreted rapidly via faeces and urine. The quail excreted about 20% 2 hours after administration. The rate of excretion on day 1 was 67.03% in the female and 71% in the male. The rate of excretion in the first 6 days exceeded 78%. The fowl excreted 5 and 14% in the female and male, respectively, in the first 2 hours. The amount excreted on day 1 was 53.26% in the female and 65.84% in the male. The amount excreted in the first 6 days was 68.32 and 75.21% in the male and female, respectively. After 6 days the viscera were autoradiographed. Sulphur-35-dimehypo was found to be distributed mainly in the intestines, with some in the muscle and none in the fat. Thus, it does not appear to accumulate in either the fowl or the quail.

TABLE IV. PARTITION COEFFICIENTS OF ^{35}S-DIMEHYPO, ^{14}C-BHC AND ^{14}C-FENVALERATE IN THE CAPRYLALCOHOL–WATER SYSTEM

Pesticide	Concentration of radioactive labelled compounds (mg/mL)	Concentration of radioactivity after equilibrium		Distribution coefficient		
		Caprylalcohol (counts·min^{-1}·0.5 mL^{-1})	Water (counts·min^{-1}·0.5 mL^{-1})	Caprylalcohol (counts/min)	Water (counts/min)	
				Measured		Average
^{35}S-dimehypo	0.15	40	16 864	0.0024		0.0016
	1.65	21	16 749	0.0013		
	15.15	17	16 982	0.0010		
^{14}C-BHC	0.26	3 227	10	323		342
	0.51	3 295	11	300		
	1.02	3 217	8	402		
^{14}C-fenvalerate	0.22	12 958	22	589		506
	0.65	12 585	25	503		
	1.07	12 732	29	439		

9. DISTRIBUTION COEFFICIENT OF ^{35}S-DIMEHYPO IN CAPRYLALCOHOL-WATER

9.1. Materials and methods

Sulphur-35-dimehypo, ^{14}C-BHC and ^{14}C-fenvalerate of an equivalent specific activity were prepared. One millilitre of the three different concentrations of each compound was placed in a test tube and 5 mL caprylalcohol and double distilled water were added. The test tube was stoppered and vigorously shaken at 15°C for two 5 minute periods, 5 minutes apart. The mixture was allowed to stand for 2 hours. A 0.5 mL aliquot from each phase was assayed by a liquid scintillation counter (Beckman Model LS 9800) to measure the radioactivity. The partition coefficient of the compounds was calculated using the equation

$$\text{Partition coefficient} = \frac{\text{radioactivity in the caprylalcohol phase}}{\text{radioactivity in the water phase}}$$

9.2. Results

The partition coefficients of ^{14}C-BHC, ^{14}C-fenvalerate and ^{35}S-dimehypo in caprylalcohol–water are given in Table IV. As the data indicate, the partition coefficient of dimehypo in the caprylalcohol–water system is much lower than that for BHC or fenvalerate, indicating that dimehypo is a relatively hydrophilic compound and, hence, is not likely to accumulate in the fat tissues.

REFERENCES

[1] ASAKAWA, Y., Plant Epidemic Prevention **30** 8 (1976) 295–296 (in Japanese).
[2] ZHANG, Heng, Environmental Protection **4** (1980) 43–44 (in Chinese).
[3] WATAMICHI, F., Pesticides **19** 2 (1972) 1–4 (in Japanese).
[4] GLEVE, A.I.G., Soil Sci. **93** 3 (1962) 211–218.
[5] HARRIS, C.L., Weeds **14** 3 (1966).
[6] CHEN Zuyi, et al., Application of Atomic Energy in Agriculture **3** (1980) 34–39 (in Chinese).
[7] HELLING, C.S., Science **162** (1963) 562–563.
[8] HELLING, C.S., Soil Sci. **35** (1971) 732–743.
[9] CHEN Xichao, et al., Sericultural Science **11** 3 (1985) 167–171 (in Chinese).
[10] CHEN Zuyi, et al., Rural Ecology, China **4** (1985) 42–43 (in Chinese).

LOSSES OF PESTICIDES FROM AGRICULTURE

J.K. KREUGER, N. BRINK
Department of Soil Sciences,
Swedish University of Agricultural Sciences,
Uppsala, Sweden

Abstract

LOSSES OF PESTICIDES FROM AGRICULTURE.

Leaching of the phenoxy acid herbicides dichlorprop, 2-(2,4-dichlorophenoxy)propionic acid, and MCPA, (4-chloro-2-methylphenoxy)acetic acid, through natural field soils with drainage water was examined after spraying two soil types (sand and clay) in late autumn and early summer. Phenoxy acids were detected in drainage water after all four spray treatments. The highest concentrations (15 to 23 µg/L) were found in drainage water from the sandy soil when sprayed under unfavourable conditions for microbial degradation. Despite more favourable conditions for degradation during the summer, small amounts of phenoxy acids were also detected in drainage water shortly after spraying (4 to 9 days); from the clay soil, dichlorprop was detectable for 2 weeks. Between June 1985 and September 1987 a total of 258 water samples were taken in streams from May to September and analysed for 90 pesticides. Seventeen compounds were identified, including ten herbicides, two fungicides and five insecticides. The most frequently found pesticides were the phenoxy acids dichlorprop and MCPA, with the highest concentrations at the time of spraying (May to June), but detectable amounts were still found in the off-spraying season. Throughout the 3 years, positive samples of one or several compounds of phenoxy acids occurred in 37% of the water samples taken in May, 78% in June, 57% in July, 24% in August and 18% in September. The maximum measured concentration of total content of phenoxy acids in one single stream was 25 µg/L in June 1985. Along with the phenoxy acids, the herbicide atrazine was found in some streams over the whole sampling season. In watersheds where only smaller parts of the area are devoted to agricultural production, no pesticides were found, or only small amounts were detected on single occasions. These investigations indicate that under certain conditions pesticide residues arising from normal agricultural use may contribute to diffuse (non-point) pollution of the aquatic environment.

1. INTRODUCTION

Phenoxy acid herbicides have long been the most frequently used pesticide group in Swedish agriculture. During 1986 sales amounted to 3000 tonnes active ingredient (a.i.), which is 50% of the total pesticide use in agriculture. The normal dose varies between 0.5 and 3.0 kg a.i./ha. Phenoxy acids are quite mobile compounds. The relative mobility of the phenoxy acid compound MCPA is ranged in class 4 on a scale from 1 to 5, with most mobile compounds in class 5 [1]. In laboratory experiments the half-lives of dichlorprop and MCPA in soil were determined as 5 to 14 days [2–6]. The relatively short degradation time, together with a normally small precipitation surplus, have lead to the assessment that the risk of phenoxy acid leaching is negligible under normal field conditions. In some reports, however, emphasis has been placed on the importance of more thorough studies of the risk of phenoxy acid leaching under field conditions with respect to their widespread use [7–9].

Over the past 20 to 30 years a considerable amount of research has been carried out on the effects of pesticide use on operator safety, food (residues) and non-target species. However, little work has been done on the effects of pesticides on the aquatic environment from normal agricultural use. Pesticide residues arising from such use have been found in the United States of America, both in shallow aquifers and in surface waters [10–15]. However, in most countries there is a lack of adequate valid comparative and quantitative data.

Swedish climatic conditions are characterized by a short (6 to 8 months) and relatively cold (3 to 17°C) vegetation period, quite often with a precipitation surplus, especially in the western parts of the country.

Surface water is estimated as supplying 50% of the Swedish drinking water needs. There are also thousands of horticulturists who are dependent upon pure irrigation water for their sensitive greenhouse cultures. An investigation carried out during the summer of 1987 among 80 growers using surface water, groundwater, municipal water or rain water showed that one-third had severe problems with their irrigation water when tested on cucumbers and tomatoes. Only one-third of the waters investigated were free from the growth of inhibiting substances [16].

The purpose of this work was: (1) to investigate the leaching of the phenoxy acid herbicides dichlorprop and MCPA in two soil types under natural field conditions, and (2) to determine the degree and extent of contamination by pesticides in stream water.

2. MATERIALS AND METHODS

2.1. Field plot study

The displacement of the phenoxy acid herbicides dichlorprop, 2-(2,4-dichlorophenoxy)propionic acid, and MCPA, (4-chloro-2-methylphenoxy)acetic acid, through a natural field soil was examined in field plots at three different locations normally used for the study of plant nutrient leaching. The techniques and procedures used have been described elsewhere [17–19] and are only briefly reviewed here.

TABLE I. SOIL CHARACTERISTICS AND TIME OF SPRAYING AT THE EXPERIMENTAL AREAS (mean of samples from all plots in per cent)

Depth (cm)	Clay <2 μm	Silt 2-60 μm	Sand >60 μm	Organic matter	pH
Field 1, spraying 24 October					
0-20	45.2	37.6	17.2	4.0	7.7
20-40	54.5	35.0	10.5	2.1	7.3
40-60	59.5	32.8	7.7	0.8	7.3
60-80	61.5	31.2	7.3	0.6	7.4
80-100	64.1	30.3	5.6	0.4	7.6
Field 2, spraying 8 November					
0-30	8.4	8.7	82.9	2.7	6.5
30-60	7.4	19.5	73.1	0.7	6.7
60-90	7.4	8.7	83.9	0.0	—
Field 3, spraying 8 June					
0-30	10.1	10.4	79.5	2.5	6.5
30-60	7.4	6.7	85.9	0.4	6.8
60-90	7.4	5.6	87.0	0.2	7.2
90-120	56.2	31.4	12.4	1.0	7.9

The experimental fields consist of six to eight plots varying between 1600 and 4000 m^2 each. The soil characteristics for the experimental areas are given in Table I. A separate subsurface drainage system runs from each plot to an underground measuring station for continual registration of the discharge. During the discharge periods water samples were collected weekly. In addition, the top metre of soil, stratified into different layers, was sampled on various occasions with the aim of following the movement of the herbicides through the soil.

Dichlorprop and MCPA were sprayed in late autumn on the soil surface (Field 1 clay, and Field 2 sand) with 2.0 kg a.i./ha and in early summer on a barley crop (Field 3 sand) with 1.5 kg a.i./ha. Half of the plots at each site were sprayed with dichlorprop and the other half with MCPA.

Measurements were made of the precipitation at all three sites. During the summer the evaporation from a free water surface was measured to give an indication of the water availability in the soil.

2.2. Field study

Leaching of the phenoxy acid herbicides dichlorprop (1.2 kg a.i./ha) and MCPA (0.6 kg a.i./ha) was studied in a clay soil after spring application (25 April) on a winter wheat crop in the south of Sweden. The size of the field was 25 ha, with no surface runoff inlets. It was systematically drained and had an underground well which collected all the water from the field before it entered an adjacent river (included in the investigation of Swedish stream waters (see Section 2.3)).

Water samples were collected in the well 1 week before spraying and then on days 4, 9, 18, 37, 75, 79, 93, 107 and 123 after spraying.

2.3. Surface water monitoring

Monitoring of Swedish stream waters started in 1985 with water sampling in seven streams in June and July. In August two of these streams were sampled. In 1986, 18 stream waters were investigated once a month from May to September; in one case they were examined in October. An additional water course was sampled once or twice a month for a whole year. In 1987 water samples were collected from 29 stream waters once or twice a month from April to May or June to September. A total of 258 water samples were collected in glass bottles. In June 1987 four duplicate samples were collected concomitantly to validate the collection and analytical techniques. Extra water samples were also collected in two streams which had only forest land in the drainage area.

In August 1986 sediments were collected from the five streams which had the most elevated pesticide content in the water samples at a depth of a few centimetres.

The watersheds cover between 2 and 4000 km^2, with eight watersheds that are smaller than 100 km^2, six watersheds in each of the groups that are 100 to 250, 250 to 500 and 500 to 1000 km^2 and three watersheds that are larger than 1000 km^2. The agricultural land varies between 13 and 90% within the watersheds.

2.4. Analytical methods

The analyses were performed at the National Laboratory for Agricultural Chemistry in Uppsala.

Unfiltered surface water samples were analysed for 90 pesticides [20] registered in Sweden in two separate screening procedures. Drainage water samples were only analysed for phenoxy acid compounds.

Phenoxy acids and the related compounds, which might be present as esters or as conjugates, were hydrolysed with alkali for 1 hour at 100°C. After acidification, the acids were extracted with dichloromethane. Extractive alkylation with pentafluorobenzyl-bromide and gas chromatography were carried out according to Ref. [21]. Verification was done with gas chromatography and mass spectrometry

TABLE II. PESTICIDE CONCENTRATIONS IN DRAINAGE WATER FROM FIELD PLOTS

Time of spraying	Pesticide	Concentration (µg/L)	1st trace (d)	Duration[a] (d)	Losses (%)
Clay					
Autumn (Field 1)	MCPA	<0.1	—	—	0
	Dichlorprop	0.2–2.5	47	29	0.06
Spring	MCPA	0.05	107	1	>0
	Dichlorprop	0.1–0.6	4	14	0.06
Sand					
Autumn (Field 2)	MCPA	1–15	41	82	0.4
	Dichlorprop	1–23	41	102	0.9
Summer (Field 3)	MCPA	0.3	9	1	>0
	Dichlorprop	0.3	9	1	>0

[a] Minimum duration.

detection. When drainage water from Field 2 was analysed the detection limit was 1.0 µg/L; on all other occasions the detection limit was 0.05 to 0.1 µg/L.

The soil samples analysed for phenoxy acid compounds were, after extraction with methanol and ammonia, essentially treated in the same way as the water samples; the detection limit was 0.001 to 0.01 mg/kg.

The semi-polar and non-polar pesticides were extracted from the water with dichloromethane. Hydrophobic gel permeation clean up and capillary column gas chromatography with selective detectors were carried out according to Ref. [20]. Verification was done with gas chromatography and mass spectrometry detection; the detection limit was 0.1 to 1.0 µg/L.

The new herbicides, chlorsulfuron and metsulfuron (0.004 kg a.i./ha), which to some extent are replacing the phenoxy acid herbicides, were not included in this study owing to the lack of adequate analytical procedures.

3. RESULTS

3.1. Field plot study and field study

Phenoxy acids were detected in drainage water after all four spray treatments, although the concentrations and duration varied greatly (Table II). The highest concentrations were found in drainage water from the sandy soil when sprayed under

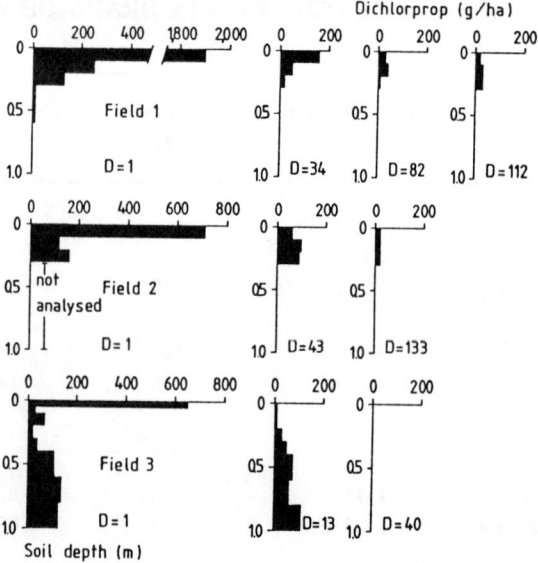

FIG. 1. *Dichlorprop content in soil from the field plots taken on different occasions after spraying (D = day).*

unfavourable conditions for microbial degradation. Despite more favourable conditions for degradation during spring and early summer, small amounts of phenoxy acids were also detected in drainage water shortly after spraying (4 to 9 days); from the clay soil, dichlorprop was detectable for 2 weeks.

Soil samples taken before spraying were clear of phenoxy acid residues. In soil samples taken from the sandy soil sprayed in early summer (Field 3) (Fig. 1), detectable amounts of MCPA and dichlorprop were found down to a depth of 1 m the day after spraying. Twelve days later, no MCPA was found in the soil profile below the 10 cm depth in plots sprayed with MCPA. However, in the adjacent plots, which were sprayed with dichlorprop, MCPA along with dichlorprop were found from the 20 to 30 cm down to the 1 metre depth, with a peak concentration in the 40 to 60 cm layer. This lateral transportation was confirmed by analysis of the drainage water samples, where MCPA was detected in water coming from the plots sprayed with dichlorprop. Most of the drainage flow on that occasion bypassed the drainage system because of the low groundwater levels.

3.2. Surface water monitoring

Seventeen compounds were identified in stream water, including ten herbicides, two fungicides and five insecticides (Table III). The most frequently found pesticides were MCPA and dichlorprop, with the highest concentrations at the time

TABLE III. MAXIMUM PESTICIDE CONCENTRATIONS AND NUMBER OF POSITIVE SAMPLES IN SWEDISH STREAM WATERS, 1985–1987

Pesticides	Maximum concentration (μg/L)	Positive samples						
		Apr.	May	Jun.	Jul.	Aug.	Sep.	Oct.
Herbicides								
Atrazine	6.0	0	5	9	14	14	13	2
Bentazone	0.5	NM	NM	NM	NM	18	18	NM
Cyanazine	0.7	0	0	10	3	1	1	0
2,4-D	0.9	0	1	3	0	0	1	0
Dichlorprop	16.0	0	14	44	24	7	5	0
MCPA	8.0	0	12	45	32	7	3	0
Mecoprop	6.0	0	13	16	19	7	3	0
Metazachlor	7.0	0	0	3	2	0	8	0
Simazine	1.1	0	0	0	1	1	1	0
Terbuthylazine	0.7	0	0	1	1	0	0	0
Fungicides								
Metalaxyl	1.3	0	0	1	2	1	0	0
Propiconazole	1.2	0	0	2	1	0	0	0
Insecticides								
Endosulfan	0.1	1	0	0	0	0	0	0
Fenitrothion	0.1	0	0	1	1	0	0	0
Lindane	0.6	0	2	2	2	2	0	0
Permethrin	0.6	0	0	0	0	1	0	0
Pirimicarb	3.7	1	0	1	2	0	1	0
No. of samples		4	42	58	56	48	43	2
Maximum concentration[a] (μg/L)		4	14	25	5	4	10	6

[a] Maximum total concentration in one single stream.
NM = not measured.

of spraying (May to June), but detectable amounts were still found in the off-spraying season. Throughout the 3 years positive samples of one or several compounds of phenoxy acids occurred in 37% of the water samples taken in May, 78% in June, 57% in July, 24% in August and 18% in September. The maximum concentration of total content of phenoxy acids measured in one single stream was 25 µg/L. The maximum mean concentration in one stream was 3.5 µg/L, which was reached during the sampling season May to September 1987.

Along with the phenoxy acids, the herbicide atrazine was found in some streams over the whole sampling season May to September, with a maximum mean concentration of 0.5 µg/L in one stream in 1986. In the stream sampled for a whole year, detectable amounts of different pesticides were found from May to November. Water samples taken in October and November contained atrazine (0.4 to 0.6 µg/L).

The highest concentrations of pesticides were detected in streams with the smallest stream drainage areas (Table IV). In large watersheds or watersheds with only smaller parts of the area devoted to agricultural production, no pesticides were found, or only small amounts were detected on single occasions. In water samples coming from the two forest streams investigated, no pesticides were detected on any occasion, indicating that aerial deposition was not a contributing source.

The analytical results of the four duplicate samples collected in June 1987 agreed very well, with the difference between the equivalent samples varying between 0 and 20%. In the sediment samples, the only pesticide found was the insecticide DDT, the sale of which has been forbidden in Sweden since 1970. It was detected in the sediments of all five streams at concentrations varying between 2 and 17 µg/kg.

4. DISCUSSION

Most of the pesticides currently used were thought to remain in the top soil environment, in part because of their relatively rapid degradation and the tendency for many pesticides to attach to minerals and organic matter. However, recent investigations have shown that preferential flow of water and solutes through macropores and other discontinuities in the soils, even under unsaturated conditions, plays a major role in the leaching of pesticides [22].

In the literature there are very few investigations where the mobility of compounds has been followed by immediate soil sampling. An exception is an experiment with the nematocide DBCP, which was detected at a depth of 90 cm after 4 hours; after 4 days concentrations were detected at 120 cm [23]. In another field experiment with the herbicide metribuzin concentrations were found at a depth of 60 cm the day after spraying and at 100 cm after 2 days [24].

TABLE IV. MAXIMUM PESTICIDE CONCENTRATION IN STREAM WATERS (values in µg/L)

Pesticide	Stream drainage area (km^2)				
	<100	100–250	250–500	500–1000	>1000
Herbicides					
Atrazine	6.0	2.0	0.3	0.05	ND
Bentazone	0.5	0.3	0.4	ND	0.05
Cyanazine	0.6	0.7	0.4	ND	ND
2,4-D	0.6	0.9	0.4	ND	ND
Dichlorprop	16.0	5.0	8.9	0.3	0.2
MCPA	8.0	3.0	2.0	0.7	1.0
Mecoprop	1.0	6.0	3.8	0.2	0.2
Metazachlor	7.0	0.8	0.2	ND	ND
Simazine	1.1	ND	0.3	ND	ND
Terbuthylazine	0.7	0.2	ND	ND	ND
Fungicides					
Metalaxyl	1.3	ND	ND	ND	ND
Propiconazole	1.2	0.5	ND	ND	ND
Insecticides					
Endosulfan	ND	ND	0.1	ND	ND
Fenitrothion	0.1	ND	0.1	ND	ND
Lindane	ND	0.6	ND	ND	ND
Permethrin	0.6	ND	ND	ND	ND
Pirimicarb	3.7	0.1	0.2	ND	ND

ND = not detected.

These investigations indicate that under certain conditions pesticide residues arising from normal agricultural use may contribute to diffuse (non-point) pollution of the aquatic environment.

The possible consequences of pesticides in stream water are, for example, an impact on water living organisms, irrigation water quality and drinking water quality. Insecticides such as endosulfan, lindane and permethrin are toxic to fish and other water living organisms. For several months the findings of lindane in this investigation exceeded the United States Environmental Protection Agency (EPA) criteria set for fresh water aquatic life (0.01 µg/L). To protect aquatic life the International Joint Commission's objective for endosulfan is 0.003 µg/L [11], which is far below the detection limit of 0.1 µg/L used in this investigation. Herbicides can also influence aquatic life. Atrazine is an inhibitor of photosynthesis and can, at concentrations of 1 to 5 µg/L, affect the photosynthesis of phytoplankton [25]. At an atrazine concentration of 20 µg/L phytoplankton growth was depressed, followed by successional changes leading to the establishment of species of phytoplankton more resistant to inhibition by atrazine [25].

Stream water with herbicides at the concentrations found in this investigation may have injurious effects if used to water greenhouse cultures. Studies have shown that phenoxy acids can disturb the growth of sensitive cultures at concentrations of 2 to 10 µg/L [26]. This has also been shown for atrazine and simazine at 0.4 to 0.6 µg/L [27].

Much attention has focused on pesticides in drinking water. At present, there are no guidelines in Sweden for pesticides in drinking water. The EPA has proposed maximum contaminant levels for about 20 synthetic organic chemicals, including two which have been found in this study: 2,4-D (70 µg/L) and lindane (0.2 µg/L). The World Health Organization has also set guidelines for some pesticides; for atrazine the provisional guideline is 2.0 µg/L. Within the European Economic Community, the proposed guidelines are set to the detection limit, i.e. 0.1 µg/L for a single pesticide and a maximum total pesticide content of 0.5 µg/L.

REFERENCES

[1] HELLING, C.S., KEARNEY, P.C., ALEXANDER, M., Behaviour of pesticides in soils, Adv. Agron. **23** (1971) 147.

[2] ALTOM, J.D., STRITZKE, J.F., Degradation of Dicamba, Picloram and four phenoxy herbicides in soils, Weed Sci. **21** (1973) 556.

[3] SMITH, A.E., Relative persistence of di- and tri-chlorophenoxyalkanoic acid herbicides in Saskatchewan soils, Weed Res. **18** (1978) 275.

[4] MOREALE, A., VAN BLADEL, R., Adsorption, dégradation et mouvement du 2,4,5-T, MCPA et carbofuran en colonne de sol homogène, Med Fac. Landbouww., Rijksuniv. Gent **46** 1 (1981) 281.

[5] SMITH, E.A., HAYDEN, B.J., Relative persistence of MCPA, MCPB and mecoprop in Saskatchewan soils, and the identification of MCPA in MCPB-treated soils, Weed Res. **21** (1981) 179.

[6] SMITH, A.E., Soil persistence studies with ^{14}C-MCPA in combination with other herbicides and pesticides, Weed Res. **22** (1982) 137.

[7] SCHMIDT, H., BEITZ, H., Erkenntnisse zum Eindringen von Pflanzenschutzmitteln in das Grundwasser and daraus abzuleitende Schutzmaßnahmen, Nachrichtenbl. Pfl.schutz DDR **34** (1980) 146.

[8] ORGANIZATION FOR ECONOMIC CO-OPERATION AND DEVELOPMENT, "Diffuse sources of agricultural pollution: Pesticides, Problems posed by their residues in the fresh water environment", Water Management Policy Group, Rep. ENV/WAT/82.1 (2nd Rev.), OECD, Paris (1983) 38.

[9] HELWEG, A., A Description of the Leaching of Pesticides in Soils: A Report of the Problem with Proposals for Measurements within this Field, Statens planteavlsforsög, Denmark (1984) 3 (in Danish).

[10] RICHARD, J.J., JUNK, G.A., AVERY, M.J., NEHRING, N.L., FRITZ, J.S., SVEC, H.J., Analysis of various Iowa waters for selected pesticides: Atrazine, DDE and dieldrin — 1974, Pestic. Monit. J. **9** (1975) 117.

[11] FRANK, R., BRAUN, H.E., VAN HOVE HOLDRINET, M., SIRONS, G.J., RIPLEY, B.D., Agriculture and water quality in the Canadian Great Lakes Basin. V. Pesticide use in 11 agricultural watersheds and presence in stream water, 1975-1977, J. Environ. Qual. **11** 3 (1982) 497.

[12] BAKER, D.B., Studies of Sediment, Nutrient and Pesticide Loading in Selected Lake Erie and Lake Ontario Tributaries. Part IV. Pesticide Concentrations and Loading in Selected Lake Erie Tributaries — 1982, Draft Final Report, United States Environmental Protection Agency, Washington, DC (1983).

[13] GLOTFELTY, D.E., TAYLOR, A.W., ISENSEE, A.R., JERSEY, J., GLENN, S., Atrazine and simazine movement to the Wye River Estuary, J. Environ. Qual. **13** 1 (1984) 115.

[14] COHEN, S.Z., CREEGER, S.M., CARSEL, R.F., ENFIELD, C.G., "Potential pesticide contamination of groundwater from agricultural uses", Treatment and Disposal of Pesticide Wastes (KRUEGER, R.F., SEIBER, J.N., Eds), ACS Symposium Series 259 (1984) 297.

[15] HALLBERG, G.R., "Agrochemicals and water quality", Colloquium on Agrochemical Management and Water Quality, Board on Agriculture, National Academy of Sciences/National Research Council, Washington, DC (1986).

[16] WIBRANT, B., LUNDIN, P., JÖNSSON, K.O., Lantbruksnämnden, Helsingborg, Sweden, personal communication, 1987.

[17] BRINK, N., "Factors affecting mass transport from farmland", Proc. Int. Sem. on Land Drainage (SAAVALAINEN, J., VAKKILAINEN, P., Eds), Helsinki University of Technology, Finland (1986) 456.

[18] KREUGER, J., Mobility of MCPA and dichlorprop, Ekohydrologi 19 (1985) 55 (in Swedish, with English summary).

[19] KREUGER, J., Mobility of MCPA and dichlorprop in a sandy soil, Ekohydrologi 20 (1985) 3 (in Swedish, with English summary).

[20] ANDERSSON, A., OHLIN, B., A capillary gas chromatographic multiresidue method for determination of pesticides in fruits and vegetables, Vår Föda **38** Suppl. 2 (1986) 79.

[21] ÅKERBLOM, M., KOLMODIN-HEDMAN, B., HÖGLUND, S., "Studies of occupational exposure to phenoxy acid herbicides", IUPAC Pesticide Chemistry, Human Welfare and the Environment (MIYAMOTO et al., Eds), Pergamon Press, London (1983) 227.

[22] JURY, W.A., ELABD, H., CLENDENING, D., RESKETO, M., "Evaluation of pesticide transport screening models under field conditions", Evaluation of Pesticides in

Ground Water (GARNER, W.Y., HONEYCUTT, R.C., NIGG, H.N., Eds), ACS Symposium Series 315 (1986) 384.

[23] BIGGAR, J.W., NIELSEN, D.R., TILLOTSON, W.R., "Movement of DBCP in laboratory soil columns and field soils to groundwater", Impact of Agricultural Activities on Groundwater, Int. Symp. IAH, Prague (1982) 47.

[24] LAFLEUR, K.S., Loss of pesticides from Congaree sandy loam with time: Characterization, Soil Sci. **130** (1980) 83.

[25] DENOYELLES, F., KETTLE, W.D., SINN, D.E., The responses of plankton communities in experimental ponds to atrazine, the most heavily used pesticide in the United States, Ecology **63** (1982) 1285.

[26] SOLYOM, P., Aspects on Irrigation Water Quality Demands, Allmänt 84, Swedish University of Agricultural Sciences, Uppsala (1986) (in Swedish).

[27] NORWEGIAN PLANT PROTECTION INSTITUTE (Statens plantevern), Pesticides in Surface Water and Groundwater, GEFO, Oslo (1987) (in Norwegian).

IAEA-SM-297/27

CONTROLLED RELEASE INSECTICIDE FORMULATIONS FOR TROPICAL APPLICATION

L. VOLLNER, A. GHODS-ESPHAHANI
FAO/IAEA Agrochemicals and Residues Unit,
IAEA Seibersdorf Laboratory,
Vienna

Abstract

CONTROLLED RELEASE INSECTICIDE FORMULATIONS FOR TROPICAL APPLICATION.
 To increase the period of residual activity of chlorinated insecticides, widely used in tsetse fly eradication campaigns, slow release formulations were developed and tested. The residual activity of DDT in the new formulation was extended to 30 days, in contrast to 5 days in commercial formulations. Similar results were obtained for dieldrin, where the half-life was extended from 1/2 day to 15 days. In the case of endosulfan, the half-life was extended from 3-7 days to 21-28 days. Bioassay of the new formulations showed that their biological activity was sufficiently high to kill tsetse flies.

1. INTRODUCTION

 In contrast to certain restrictions on the use of some chlorinated hydrocarbons (mainly insecticides) in industrialized countries, production and application in developing countries have increased. Recent studies made by the World Health Organization [1] and the Economic and Social Commission for Asia and the Pacific [2] confirm this fact and show the need for extended use of insecticides such as DDT, dieldrin, lindane, aldrin and endosulfan.
 The main users of these chemicals are tropical countries, where mosquitoes, ticks, termites, tsetse flies, etc. cause tremendous problems. On the other hand, excessive use of such chemicals may create severe hazards to the environment [3]. After application in agricultural areas, appreciable amounts of residue in feed and food can be detected. This presentation will focus on insecticides to be used for the control of tsetse flies (*Glossina palpalis palpalis*). Different species occur in more than 35 African countries in an area of about 9 million km², or 42 % of the land area of the continent. The control of tsetse flies in these countries is of vital importance as they are the vectors of trypanosomiasis and prevent the use of large areas with a high potential for agricultural development.
 Over the years, substitute compounds for chlorinated insecticides have been applied and despite the good results achieved in laboratory tests, none have shown

the economy and efficacy of DDT, dieldrin or endosulfan. Tests with synthetic pyrethroids proved promising, but owing to the high costs involved widespread use has not yet materialized.

Against this background, development of new formulations of chemicals to reduce the environmental problems and costs (by extending the effective period of the biologically active materials) should seriously be considered.

Emphasis should be placed on low cost formulation agents. These should be non-hazardous to man and the environment and readily biodegradable. The materials should protect the active ingredients (a.i.) against evaporation, ultraviolet (UV) degradation and washing out by heavy rains, and at the same time should retain activity, as needed.

In 1982, in order to assist developing countries in this field, the Joint FAO/IAEA Division of Isotope and Radiation Applications of Atomic Energy for Food and Agricultural Development, in co-operation with a number of laboratories in developed and developing countries, initiated research on the controlled release formulation of biologically active chemicals to reduce the residues and to increase pesticide efficacy [4].

This paper presents some properties of the new formulations of endosulfan and other important chlorinated insecticides such as DDT, lindane and dieldrin.

2. MATERIALS AND METHODS

2.1. Preparation of alginate and Natrosol formulations

Algin is a linear polysaccharide derived from brown seaweed. Sodium alginate (obtained from Fluka AG, Buchs, Switzerland), the most common algin, is water soluble with unique thickening, suspending, emulsifying, stabilizing, water holding and gel forming properties. For the preparation of water insoluble films or beads, sodium alginate formulations were treated with $CaCl_2$ solutions to replace Na ions with Ca ions, which cross-link the linear polysaccharide chains.

Natrosol is a non-ionic, water soluble polymeric hydroxyethyl ether derived from cellulose. Natrosol (obtained from Neuber Chemie, Vienna, Austria) dissolves readily in water and is used to produce solutions with a wide range of viscosity. Natrosol films can be made water insoluble by reacting with resin and catalyst. For our experiments we used the following formulation procedure: 0.5 g of NH_4Cl was dissolved in 50 mL of water and heated to 50°C; 0.5 g of Natrosol was then added slowly and mixed. After complete solution, pesticides, dissolved in organic solvents, were added and polymerization was initiated by adding 1.8 mL of melamine-formaldehyde 56% (Reichhold Chemie, Vienna, Austria).

Analytical grade insecticides were supplied by Supelco, Bellefonte, Pennsylvania, United States of America, and technical grade endosulfan and three different formulations (thiodan) were obtained from Hoechst AG, Frankfurt, Federal Republic of Germany.

2.2. Residue analyses

Analyses were carried out by gas chromatography (Packard model 433), using packed and capillary columns and electron capture detection. Further analysis was conducted by gas chromatography–mass spectrometry (HP 5959), with a quadrupole system. Sample combustion was done by using a Packard oxidizer (model 306) as well as a liquid scintillation counter (Nuclear Enterprises LSC-2) combined with a rate meter (SR 7).

2.3. Evaporation

The films were placed into evaporation equipment at a temperature of $50 \pm 3°C$ and an air flow of 10 L/h, or exposed to sunlight outdoors to simulate the evaporation of a.i. under dry conditons.

2.4. Release experiments

To simulate heavy rainfall the films were submerged in 500 mL of demineralized water. The increase in the pesticide concentration in solution was measured initially every 15 minutes for the first 4 hours, subsequently hourly for the next 4 hours and daily for 7 days. For release tests, we used water insoluble alginate formulations.

2.5. UV irradiation

Films were exposed to a UV lamp (Philips HPK 125 kW) at a distance of 10 cm to simulate the high energy solar radiation in dry environments using wave lengths between 280 and 600 nm.

2.6. Biological activity

The films were tested on tsetse flies obtained from the Agency's Entomology Unit. The flies were kept in PVC cages in an environmental chamber (Heraeus VEPHL 5/1350) at 20°C and 80% humidity.

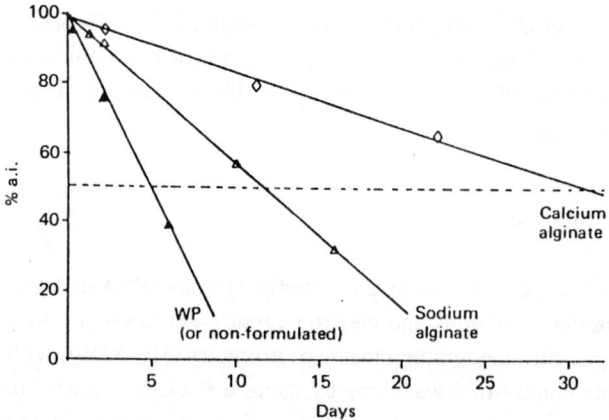

FIG. 1. *Evaporation of DDT from different formulations at a temperature of 50°C and an air flow of 10 L/h (WP = wettable powder).*

3. RESULTS AND DISCUSSION

It is known that the extreme environmental conditions in tropical and subtropical regions cause more rapid degradation, evaporation or leaching of a.i. than under temperate conditions. To improve formulations for special use, e.g. for tsetse control in dry and wet seasons, we measured the evaporation rates, the release rates into water and the photostability of the formulations. A bioassay was also conducted using tsetse flies.

3.1. DDT

Figure 1 shows the evaporation of DDT from: wettable powder, which was one of the commercial formulations; sodium alginate, which was the water soluble formulation; and the cross-linked Ca-alginate formulation, which was of the non-water soluble type. The half-life of the a.i. in the commercial formulation was 5 days under our experimental conditions (50°C continuously and an air flow of about 10 L/h). The sodium alginate formulation extends the half-life of DDT to 10 days, and Ca-alginate extends it to 30 days, which is a sixfold improvement.

3.2. Dieldrin

Similar results were obtained for dieldrin (Fig. 2). While the sodium alginate matrix extends the half-life period from 1/2 to 2 days, the water insoluble type improves the period of half-life to 15 days, which is more than 20 times longer than the half-life of the commercial formulation.

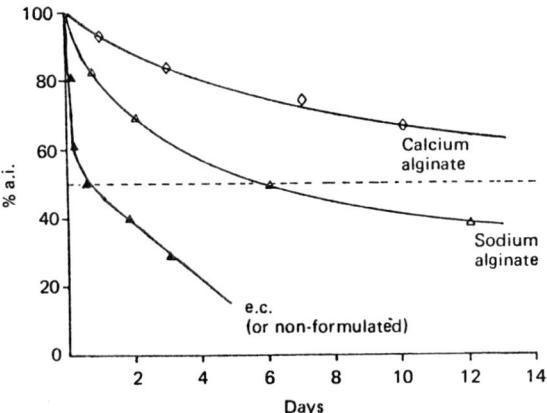

FIG. 2. Evaporation of dieldrin from different formulations at a temperature of 50°C and an air flow of 10 L/h.

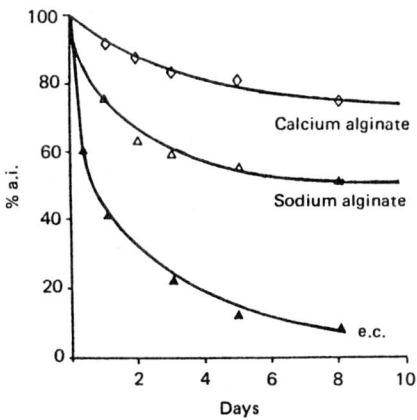

FIG. 3. Evaporation of endosulfan from different formulations at a temperature of 50°C and an air flow of 10 L/h.

3.3. Endosulfan

We investigated endosulfan, the most important insecticide currently used in the tsetse areas, more extensively. Besides comparing thiodan, a widely used formulation, with alginate formulations (Fig. 3), we also used wettable powder and dispersion materials, which are available commercially. Since technical grade insecticide is insoluble in water, such preparations are needed for large scale applications to obtain homogeneous solutions for sprays. All formulations showed improved residual properties when compared with commercial formulations.

TABLE I. EVAPORATION OF THE TWO
ENDOSULFAN ISOMERS (ALPHA AND BETA)
FROM TSETSE SCREENS AFTER A
5 WEEK EXPOSURE TIME
(field experiments in Nigeria)

	Thiodan e.c. experiments (%)		Thiodan alginate experiments (%)	
	1	2	1	2
Alpha	1.6	1.5	33.8	24.8
Beta	34.0	25.0	90.8	86.7

Small scale field tests of the different endosulfan formulations were carried out in Nigeria during a 5 week period in May and June 1985. The samples were applied to blue cotton screens, which are used after impregnation with pesticides for tsetse suppression in infested areas prior to application of the sterile insect technique [5]. The unprotected insecticides usually evaporate rapidly from the screens. The results of these experiments (see Table I) show the significant differences in the extended residual behaviour of the insecticide on targets.

3.4. Release into water

It is well known that many pesticides evaporate at a higher rate from a wet medium. In tropical areas, particularly in the rainy season, it is important to protect pesticides against washing out processes and subsequent vapour evaporation. According to the water solubility of the different materials listed below, the release rates decrease from lindane to DDT in order of succession (Fig. 4): lindane: 6000 μg/L (we included lindane, since it is still an important insecticide and is highly volatile); endosulfan: 500 μg/L; dieldrin: 200 μg/L; and DDT: 5 μg/L. The maximum accumulated concentration was reached after 1 to 2 days for DDT and 4 to 5 days for lindane, dieldrin and endosulfan; DDT was released at 13.4%, dieldrin at 7.6%, endosulfan at 65.2% and lindane at 76.5%. Release of active ingredients started again when the water was replaced.

3.5. Photostability of the formulations

We used endosulfan for the photostability experiments because of its high instability to UV light [6]. Following irradiation of the alginate and the commercial

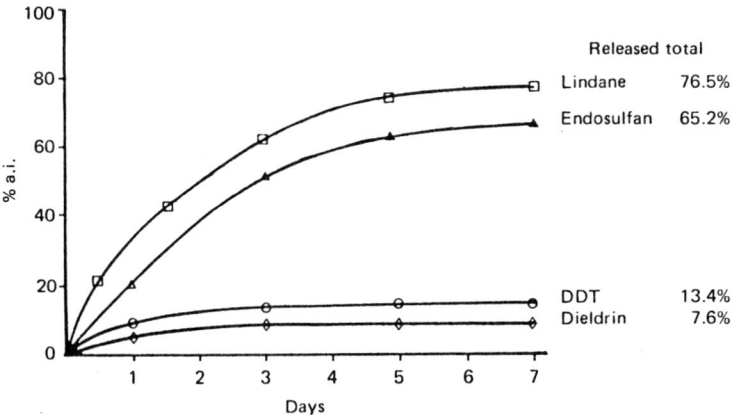

FIG. 4. Release rates of chlorinated insecticides from alginate based CR formulations.

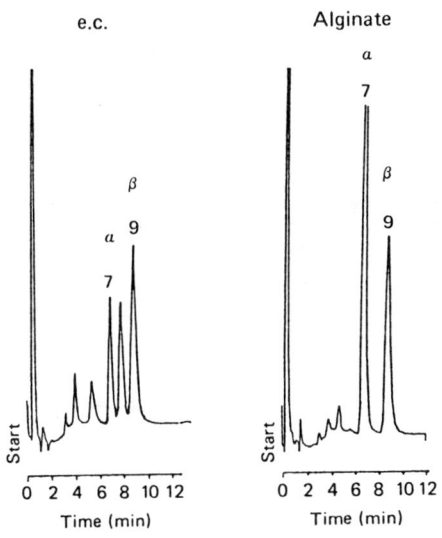

FIG. 5. UV stability of different endosulfan formulations during 4 days' exposure to sunlight (June 1985). (GC conditions: packed column: 1 m; 10% SE 30 on Chromosorb W: 220°C; ^{63}Ni-ECD: 250°C; N_2 25: mL/min. Isomer mixture of alfa- and beta-endosulfan applied: 2:1. Beta-endosulfan corresponds to 360 pg. Injected volume: 1 µL.)

e.c. formulations, it was clearly seen that decomposition of the a.i. (alpha- and beta-endosulfan) of the e.c. formulation started immediately, resulting in three different degradation products, while only traces of degradation products were formed from the alginate formulation (Fig. 5). Table II shows the quantitative data of this experiment. The more biologically active alpha-endosulfan (with a retention time of 7 min)

TABLE II. QUANTITATIVE ANALYSIS OF ENDOSULFAN AND DEGRADATION PRODUCTS IN DIFFERENT FORMULATIONS AFTER EXPOSURE TO SUNLIGHT (see Fig. 5)

Retention time of components (min)	e.c. (%)	Alginate (%)
3.4	7.2	—
5.4	7.7	—
7.0	21.7	60.4
8.0	23.0	—
9.0	40.1	39.6

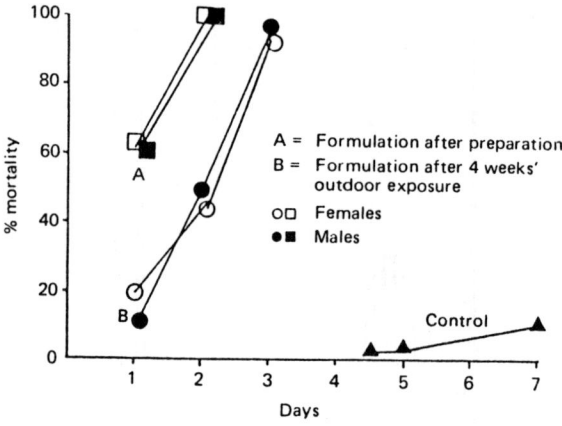

FIG. 6. Mortality of tsetse flies after 10 min exposure to CR endosulfan formulations.

decomposes at a slower rate than in e.c. formulations, thus extending the residual activity of the pesticide.

3.6. Biological activity

We tested endosulfan formulations for their biological activity against tsetse fly before and after exposure to natural environmental factors such as wind, sunshine and rain. After several weeks we again placed the strips (which carried the formula-

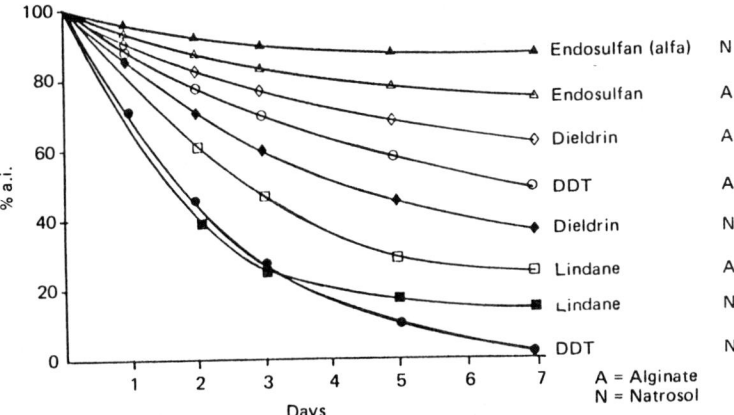

FIG. 7. Evaporation rates of chlorinated insecticides from CR formulations at a temperature of 50°C and an air flow of 10 L/h.

tions) into tsetse cages for 10 minutes (each cage contained 25 males and females). Figure 6 shows the results of these experiments. It was clear that, for example, after 4 weeks' exposure of formulations to environmental influences, the mortality of the flies was still high and reached almost 100% after 3 days. In the control experiments, mortality was only about 10% after 7 days. Formulations of other insecticides are currently being tested.

4. CONCLUSIONS

A number of chlorinated insecticide formulations were prepared. Figure 7 shows a comparison made of the evaporation curves of alginate and Natrosol formulations of the insecticides discussed. Since these biologically active compounds show different evaporation rates from two similar matrices, proper selection can be made for different applications. At a temperature of 50 ± 3°C the half-life varies from 1.5 days (lindane, DDT) to many weeks (dieldrin, endosulfan).

ACKNOWLEDGEMENTS

This work was conducted as part of a project funded by the Government of the Federal Republic of Germany, whose support we would like to acknowledge. The authors are also grateful to Hoechst AG, Frankfurt, Federal Republic of Germany, for generous cost free provision of endosulfan and other materials, and for the very useful discussions held with various staff members, especially R. Heinrich, Pesticide

Formulation Laboratory. We would also like to acknowledge Neuber Chemie and Reichhold Chemie, Vienna, Austria, for supplying samples of a series of formulating agents. We wish to thank the Gesellschaft für Strahlen- und Umweltforschung mbH, Munich, Federal Republic of Germany, for technical support, and the Entomology Unit of the Joint FAO/IAEA Programme in Vienna for performing biological tests.

REFERENCES

[1] SMITH, H., LOSSEV, O., Pesticides and Equipment Requirements for National Vector Control Programmes in Developing Countries, 1978-1984, Rep. VBC/81.4, World Health Organization, Geneva (1984).
[2] STARING, W.D.E., Pesticides, Data Collection Systems and Supply, Distribution and Use, Economic and Social Commission for Asia and the Pacific, Bangkok (1984).
[3] MATTHIESSEN, P., Environ. Pollut., Ser. B **10** (1985) 189.
[4] INTERNATIONAL ATOMIC ENERGY AGENCY/GESELLSCHAFT FÜR STRAHLEN- UND UMWELTFORSCHUNG, Joint Programme of Research to Develop and Evaluate Controlled Release Formulations of Pesticides to Reduce Residues and Increase Efficacy, Agreement of 1981.
[5] FEDERAL DEPARTMENT OF PEST CONTROL SERVICES, Biological Control of Tsetse by the Sterile Insect Technique, Kaduna, Nigeria.
[6] SCHUMACHER, G., KLEIN, W., KORTE, F., Tetrahedron Lett. **24** (1971) 2229.

Memoria encargada

EL USO DE PLAGUICIDAS EN AMERICA LATINA
Tendencias e implicaciones ambientales

M.A. CONSTENLA
Universidad de Costa Rica,
San José, Costa Rica

Abstract-Resumen

PESTICIDE USE IN LATIN AMERICA: TRENDS AND ENVIRONMENTAL IMPLICATIONS.
 The current demand for pesticides in Latin America exceeds US $2000 million in value. Up to 1990, the consumption is expected to rise annually by 7 to 12%. Use of pesticides has been fully accepted in Latin America as a readily available means of pest control. Nevertheless, even today pests are responsible for the loss of 25 to 40% of potential harvests. In banana cultivation, the cost of pesticides accounts for 30 to 35% of production costs. Latin America is in the process of changing over from persistent organochlorines to pesticides which are more sophisticated, more expensive and less persistent. The countries in the region are looking for ecologically acceptable permanent methods of pest control. Integrated pest management will reduce the risks of pesticide use by changing the systems of application. All Latin American countries have legislation on pesticide handling and use. These laws have been or are being harmonized with the International Code of Conduct on the Distribution and Use of Pesticides published by FAO. The Latin American pesticide industry is dependent on the supply of intermediate compounds and raw materials from developed countries. In addition, it has to deal with high production costs and a lack of research and development. It is far from utilizing the full installed capacity, and only an increase in commercial exchange in the region would make its position less unfavourable.

EL USO DE PLAGUICIDAS EN AMERICA LATINA: TENDENCIAS E IMPLICACIONES AMBIENTALES.
 Actualmente, la demanda de plaguicidas en América Latina se cifra en más de 2000 millones de dólares de los EE UU. Se estima que hasta 1990, su consumo aumentará en un 7 al 12% anual. El uso de plaguicidas ha sido plenamente aceptado en América Latina como el control inmediato disponible para combatir las plagas. Sin embargo, se pierden aún hoy en día del 25 al 40% de las cosechas potenciales a causa de las plagas. En el cultivo del banano el 30 al 35% del costo de la producción corresponde a los gastos en plaguicidas. América Latina se encuentra en el proceso de cambio del uso de plaguicidas organoclorados persistentes por plaguicidas más elaborados, más caros y menos persistentes. Los países de la región están a la búsqueda de métodos de control permanentes de las plagas, compatibles con la ecología. El manejo integrado de las plagas reducirá los riesgos del uso de los

plaguicidas, mediante el cambio de los sistemas de aplicación. Todos los países latinoamericanos tienen legislaciones en el manejo y en el uso de plaguicidas. Estas legislaciones han sido o están siendo armonizadas de acuerdo con el Código Internacional de Conducta para la Distribución y el Uso de Plaguicidas de la FAO. La industria latinoamericana de plaguicidas depende del suministro de compuestos intermedios y materias primas por parte de los países desarrollados. Además, tiene que afrontar costos de producción mayores y la falta de investigación y desarrollo. La industria está lejos de utilizar toda la capacidad instalada, y solamente un aumento del intercambio comercial en la región le ayudaría a ocupar una posición menos desfavorable.

1. INTRODUCCION

Después de la segunda guerra mundial se desataron desarrollos progresivamente acelerados en todo el planeta. En América Latina se creó una gran confianza en la aplicación de técnicas modernas para aumentar la productividad agrícola y pecuaria, y se adaptaron modelos de desarrollo copiando los que provenían de los países desarrollados, de otras latitudes.

Estos conceptos y modelos implementados en los países latinoamericanos han generado un complejo de distorsiones que está repercutiendo a menudo de forma negativa en el bienestar de las comunidades, repercusiones que se agudizan al no haberse contemplado el medio ambiente y su dinámica como un elemento constitutivo y necesario de la calidad de vida del ser humano [1].

Se ha enjuiciado el desarrollo agroexportador de América Latina como un desarrollo forzado bajo condiciones de intercambio desigual que supedita a los países de esta área a importar tecnologías frecuentemente contaminantes.

Sin embargo, se tiene conciencia de que el medio ambiente y el desarrollo no son realidades antagónicas.

En América Latina se está desarrollando la convicción de que el desenvolvimiento científico, tomando en cuenta los valores nacionales así como la tecnología propia y sustentado por amplias campañas educacionales, permitirá reducir los problemas ecológicos de la región.

Existe muchísima literatura sobre los plaguicidas en América Latina, sobre todo en los últimos 17 años. Considero imposible tratar de condensar en esta oportunidad todo el trabajo realizado en la región. Contrario a lo que se cree, cada país latinoamericano ejerce actividades de investigación, desarrollo y legislación a través de tres canales institucionalizados:

1) Los ministerios de agricultura y sus centros de investigación.
2) Las universidades.
3) Ministerios de salud y sus centros de investigación.

Las compañías transnacionales, a su vez, financian investigaciones bajo condiciones locales, para conocer mejor el funcionamiento de sus productos.

La necesidad del desarrollo de una agricultura más eficiente ha requerido de técnicas y tecnologías que involucran el uso de sustancias químicas, la mecanización y el mejoramiento genético de las plantas.

Es de todos bien conocido que los monocultivos y las plantaciones extensivas generan condiciones muy favorables para el desarrollo de las plagas. Esto tuvo como consecuencia la necesidad de recurrir a aumentar el uso de los agroquímicos para proteger las cosechas.

El uso de los plaguicidas en la agricultura sobrepasa el objetivo inicial de controlar las plagas. Los herbicidas son indispensables para la mecanización de los cultivos, y el empleo de defoliantes y desecantes facilita la recolección mecánica de las cosechas.

El uso de plaguicidas de origen vegetal, entre los que se cuentan las piretinas y la nicotina, data de tiempos anteriores a la revolución industrial. Luego, en tiempos de ésta, se popularizó el uso de los plaguicidas inorgánicos, algunos de los cuales son subproductos de la minería. Su bajo costo hizo que grandes cantidades de sales de arsénico, cobre, plomo y de mercurio se utilizaran sin restricción alguna en el mundo, pese a ser altamente persistentes; algunas de ellas se utilizan aún hoy día.

Los dos tipos de productos tanto de origen vegetal como de tipo inorgánico se les ha catalogado como "primera generación" para distinguirlos de la "segunda generación", o sea los productos sintéticos de la química orgánica y que se produjeron por efecto de las demandas de la segunda guerra mundial [2]. La "tercera generación" ha sido definida como la generación del uso de hormonas de insectos como insecticidas más selectivos, a prueba de desarrollo de resistencias.

¿Dónde nos encontramos en América Latina y cuáles son las tendencias futuras?

En el uso de los plaguicidas, América Latina se encuentra en la evolución de la segunda generación, caracterizada por la adopción de plaguicidas más costosos y de tecnología más reciente, que se caracterizan además por una persistencia disminuida.

Debemos tener en mente que las plantas cultivadas por el hombre pueden ser afectadas por 80 000 a 100 000 enfermedades causadas por hongos, bacterias, viruses, ricketsias, algas, microplasmas y espiroplasmas que parasitan también a las semillas y a los productos almacenados. Existen además alrededor de 30 000 especies de malas hierbas que compiten con los cultivos; 1800 de ellas son un problema serio en la agricultura. Además tenemos 1000 especies de nemátodos y 10 000 especies de insectos que parasitan a las cosechas.

Se ha estimado que en ausencia de los plaguicidas, los países latinoamericanos perderían el 50% de la cosecha de algodón.

Mucho se ha comentado en todos los foros sobre el uso abusivo de los plaguicidas en América Latina. Dos son las fuentes de este comportamiento de los agricultores: ignorancia y el deseo de lucro.

En todos los países de América Latina, en unos más y en otros menos, se han impulsado programas educacionales y servicios técnicos de extensión agrícola. Sin embargo, principalmente los agricultores pequeños y medianos continúan abusando y oponiéndose a las enseñanzas del extensionista.

2. USO DE PLAGUICIDAS EN AMERICA LATINA

2.1. Diversidad climática y geográfica

La diversidad climática y geográfica en América Latina es posiblemente única en todo el mundo. Mientras que en Africa es necesario movilizarse generalmente mil o varios miles de kilómetros para estudiar un cambio de textura en los suelos, en América Latina se pueden estudiar diez o más texturas de suelos diferentes en distancias de unos pocos cientos de kilómetros.

La diversidad climática varía desde climas similares a la tundra hasta los climas tropicales del Caribe húmedo en América Central; se tienen así condiciones climáticas que varían no sólo con la latitud sino también con la altura.

Por otra parte encontramos las condiciones climáticas de las grandes planicies, y los microclimas de los valles andinos.

Las condiciones también varían en las cordilleras, con los cultivos sembrados en fuertes pendientes en suelos altamente erosionables.

Se abre de esta manera un campo muy prometedor para la industria de los agroquímicos y para la investigación, quienes podrían proveer de soluciones específicas a problemas de la agromedicina relacionados con la gran diversidad de condiciones climáticas y geográficas en América Latina.

2.2. Demanda de plaguicidas en América Latina

En la mayoría de los países latinoamericanos no existe capacidad local de producción de ingredientes activos. En algunos pocos países se han establecido empresas nacionales e internacionales que sintetizan algunos de los plaguicidas requeridos en la región. Se importan además muchos ingredientes activos e inclusive las formulaciones terminadas desde los países más desarrollados.

Las estadísticas sobre importaciones son deficientes e incongruentes. No obstante, las cifras disponibles permiten establecer un panorama general sobre el incremento en el uso de los plaguicidas, sobre todo para fines agrícolas. En la Fig. 1 se puede observar el incremento de la importación de plaguicidas en Guatemala y en Costa Rica desde 1970 [3, 4].

El incremento fue sostenido y muy por encima de las predicciones [5] en el decenio 1970–1980, período en el cual el volumen aumentó en un 116% y su costo

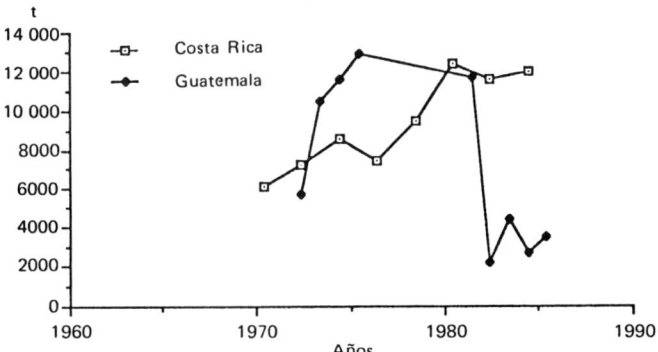

FIG. 1. *Importación de plaguicidas (ingredientes activos) en Guatemala y Costa Rica (1970–1985).*

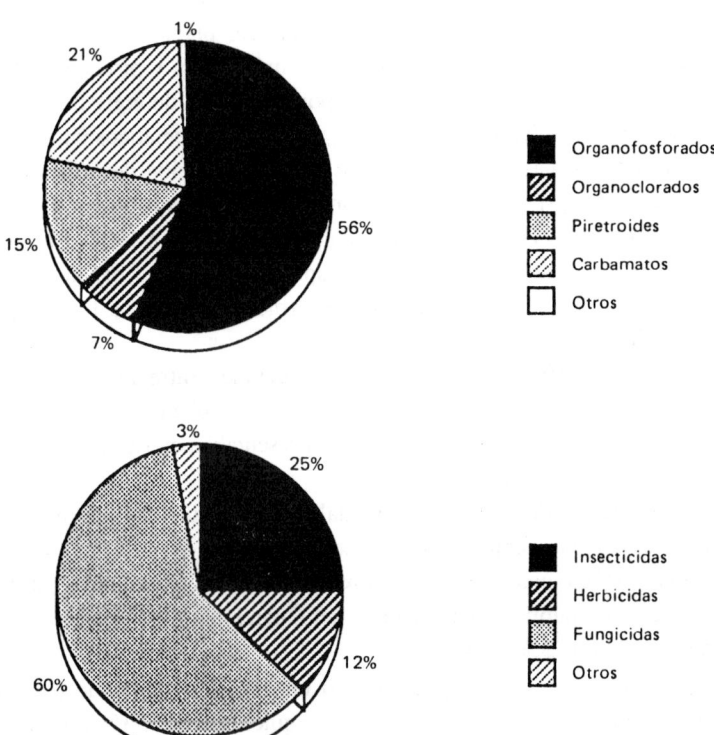

FIG. 2. *Uso de insecticidas y plaguicidas en Guatemala (costos), 1985.*

en un 550%. Lo cual, por su parte, indica un mayor uso de plaguicidas más sofisticados y por ende más caros.

En Honduras la importación de plaguicidas aumentó entre los años de 1980 a 1984 en un 34% en valor y un 35% en volumen, equivalentes a un aumento anual del 7%.

Se puede considerar que el crecimiento en el uso de plaguicidas en América Latina se mantendrá hasta el año 1990 entre un 7 y un 12% anual, respecto al volumen, si no se presentan crisis económicas globales.

Sin embargo, el consumo de plaguicidas puede incrementarse de forma inesperada, como ha sucedido recientemente en El Salvador, donde en 1985 el consumo se cifró en 7000 toneladas métricas, y en 1986 pasó a 11 000 toneladas debido al resurgimiento del algodón.

Así, las predicciones de Maltby [6] para América Central y para el año 1988, que proyectan un consumo de 30 703 t y un costo de dóls. (EE UU) 165 000 000, han sido superadas en el año 1986.

La tendencia con respecto a los tipos de plaguicidas usados se desprende también del aumento en los costos por kilogramo. En Costa Rica, el costo por kilogramo de plaguicida importado aumentó, de 1970 a 1980, de dóls. 0,90/kg a dóls. 3,10/kg. La tendencia es general en América Latina. Se están dejando de usar plaguicidas organoclorados de menor costo y se utilizan más los plaguicidas organofosforados (dos veces más caros), carbamatos (tres veces más caros) y piretroides (seis veces más caros), lo cual ha triplicado los costos por kilogramo de ingrediente activo. Sin embargo, por ser más eficientes, selectivos y menos problemáticos para al ambiente, su aplicación resulta más barata a la larga que la aplicación de organoclorados.

En Guatemala [4] los insecticidas organoclorados constituían el 49% de los insecticidas importados en 1978, el 38% en 1981 y el 7% en 1985. En 1985 la distribución de los diferentes insecticidas fue aproximadamente: organoclorados 7%; organofosforados 56%; carbamatos 21%; piretroides 15%; otros 1%, tal y como se puede ver en la Fig. 2.

También han sufrido cambios las distribuciones entre los diferentes plaguicidas: en 1978 los insecticidas constituían el 86% de los plaguicidas usados y en 1985 únicamente el 25%. En 1978 los fungicidas representaban el 3%, para pasar en 1985 al 60% del volumen comercial.

En la evaluación del impacto ambiental de los plaguicidas en Américan Latina es necesario tomar en cuenta estas tendencias.

Además, se da una reducción porcentual de los plaguicidas utilizados en programas de salud pública, por ejemplo, en Guatemala pasó del 6% del volumen importado en 1978 al 2% en 1984 [4].

En Argentina, en 1982 [7] se alcanzó un valor de ventas de plaguicidas de dóls. 123 millones. La distribución de los plaguicidas fue: 53% herbicidas; 35% insecticidas; 11% fungicidas y otros un 5%. El aporte de la industria nacional ha sido del 25% del total, y el resto, los productos importados, proviene de EE UU, Europa e Israel.

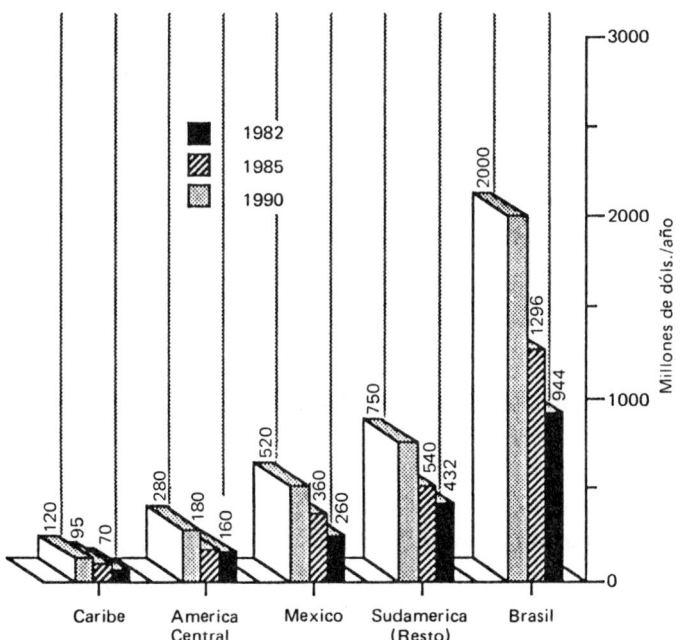

FIG. 3. *Demanda y predicciones de demanda de plaguicidas en América Latina.*

Los productos usados más importantes son:

a) Insecticidas como monocrotofos, paratión, endosulfán, malatión y piretroides.
b) Herbicidas como eptam, alaclor, glifosato, atrazina, bentazón, dalapón, picloram y dícamba.
c) Fungicidas como los ditiocarbamatos.

La demanda de plaguicidas en América Latina alcanzó en el año 1982 aproximadamente dóls. 1634 millones. El pronóstico del crecimiento del consumo es de un 9% anual entre 1985 y 1990. En la Fig. 3 se pueden apreciar estas proyecciones.

La demanda de plaguicidas en México tendrá un crecimiento estimado del 10% [7]. La demanda en Sudamérica se estima que crecerá al ritmo de un 9% anual. Brasil sustenta actualmente el 71% de la demanda y, por lo tanto, es el mayor consumidor de plaguicidas de Sudamérica y de toda América Latina. Brasil consume igual o más cantidad de plaguicidas que todo el resto de América Latina.

En América Central se espera un crecimiento promedio de un 10 al 12% anual hasta 1990. La demanda de plaguicidas en esta zona depende mucho de fenómenos conyunturales y se pueden dar aumentos muy considerables de un año a otro, así como contracciones violentas del mercado.

La mayoría de los países latinoamericanos importan productos formulados. Sin embargo, existen empresas formuladoras en casi todos los países. Brasil,

Argentina y México formulan la mayoría de los productos que requieren. Perú, Bolivia, Argentina, Brasil, Uruguay, Chile, México, Colombia y Guatemala tienen producción industrial de algunos ingredientes activos. Pero la mayoría de los países productores están produciendo por debajo de la capacidad instalada y enfrentan problemas comunes.

Con las mejoras de infraestructura en América Latina se observa una mejor distribución de proveedores de plaguicidas. El aumento de empresas agroquímicas (sin incluir los detallistas) es un indicador importante. Así, en Costa Rica, pasaron de 6 empresas en 1950 a 160 en 1983 [8].

2.3. Tipos de plaguicidas en uso, tendencias

2.3.1. Insecticidas

En Sudamérica, Brasil sostiene un 70% del mercado de insecticidas, mientras que Argentina y Colombia juntas hacen un 16%. En Brasil [6] se vendieron unas 14 500 t de ingrediente activo de organoclorados; para 1988 se espera que se reduzca el uso de estos productos a 12 000 t, las reducciones mayores serán en DDT, lindano y toxafeno y el uso de endosulfán aumentará.

En cuanto a los organoclorados, se espera que la demanda aumente en 1988 a unas 12 000 t. Los insecticidas monocrotofos, metilparatión, dimetoato y dicrotofos son los que están alcanzando mayores aumentos.

En lo que se refiere a los carbamatos, se espera que unas 2300 t de ingrediente activo se vendan en el mercado.

Respecto al uso de insecticidas piretroides, América Central y México estarán consumiendo en 1988 aproximadamente diez veces más que Brasil, país que también aumentará su consumo. Lo que pone de manifiesto la apertura de esos mercados para la introducción de nuevas tecnologías.

Como hemos visto, en el caso de América Central estas predicciones son conservadoras y han sido superadas. Pero las tendencias de uso expuestas por estos datos se mantienen.

2.3.2. Herbicidas

En Brasil los herbicidas fenólicos como el DNBP irán siendo sustituidos, mientras que los fenoxiacéticos como el 2,4-D aumentarán sus ventas hasta duplicarlas en 1988 con respecto a 1978.

Los herbicidas carbamatos se mantendrán estáticos y en algunos casos su consumo disminuirá, por ejemplo, el azulam (metilsulfanilcarbamato) y el molinate.

Los herbicidas derivados de las ureas sustituidas, como el diurón, se espera que en 1988 alcancen un 80% más del consumo que en 1978. Lo mismo se espera de los herbicidas triazínicos. La diazina, como el bromacilo y la bentazona, mantendrá su nivel de ventas.

Las amidas, como el propanilo, el alaclor y el metolaclor, seguirán aumentando su uso.

Los herbicidas toluidínicos, tan importantes para el cultivo de la soya, duplicarán sus ventas en Brasil al final del período 1978–1988.

El dalapón, el glifosato, y el picloram son otros herbicidas con tendencias a aumentar sus ventas.

Entre 1978 y 1988 el paraquat y el diquat habrán aumentado las ventas en un 300%.

2.3.3. Fungicidas

En cuanto al uso de fungicidas en Brasil, hay una tendencia cada vez mayor a disminuir el uso de los fungicidas inorgánicos, con excepción del azufre en forma de polvo mojable.

Los fungicidas clorofenólicos se mantienen estáticos en sus ventas, mientras que los ditiocarbamatos mantienen una tendencia de alza en su consumo. Otros fungicidas como el kitazin, el triadimefón y el tridemorph aumentarán todavía más su consumo.

En Argentina y en el resto de América Latina se manifiestan tendencias de uso muy similares.

2.4. Producción, distribución y registro de plaguicidas en América Latina

El uso de plaguicidas actual y el proyectado para América Latina es menor que las necesidades explicadas en términos de niveles deseables en dieta, salud y capacidad de exportación. Existe una subutilización de todo el paquete tecnológico (mecanización, fertilización, irrigación, manejo del suelo y mejoramiento genético). Llera [7] considera que el consumo de plaguicidas debería incrementarse 10 veces para lograr una disminución significativa de las pérdidas de cosechas.

La región latinoamericana tiene capacidad para cubrir las demandas de producción en cuanto a un número importante de plaguicidas, y tiene la capacidad para llenar todas las demandas de formulación. Sin embargo, existen grandes trabas al intercambio complementario entre los países.

Por esa razón, para América Latina la armonización de los requisitos para el registro y la uniformación de la etiqueta, siguiendo los lineamientos de las organizaciones internacionales como la FAO y la OMS, es una manifiesta necesidad. Así lo ha recomendado en repetidas ocasiones el Programa Hemisférico de Sanidad Vegetal.

En el campo de armonización de criterios en registro, etiquetado y control de calidad de plaguicidas, en un período de pocos años (1979 a 1983) y bajo la coordinación del Instituto Interamericano de Cooperación para la Agricultura (IICA), se ha venido configurando un conjunto de recomendaciones relativas a la armonización de los puntos citados y de la clasificación toxicológica.

En el proceso de implementación de la armonización, que se inició en 1979 en el IICA en Costa Rica, han habido reuniones posteriores en 1981 en la Florida, EE UU; también en 1981 Contadora, Panamá [9]; y en 1982, en México, se celebró la tercera Consulta Gubernamental del Uso Adecuado de Plaguicidas en América Latina y el Caribe.

En 1982 se celebró en Colombia la Reunión de Consulta para los países del área Andina [10]. En 1983, en Costa Rica, se celebró la II Consulta Intergubernamental sobre Armonización de Criterios para Registros y Etiquetado de México, Centroamérica y Panamá [11], en la que se acordó armonizar con las normas adoptadas por la región andina. Posteriormente, en la I Reunión de Consulta de los países del Caribe, estos aceptaron las normas del área andina y de América Central y México.

CUADRO I. DISTRIBUCION DEL USO DE INSECTICIDAS POR CULTIVO EN AMERICA LATINA

Cultivo	Uso (en %)			
	Argentina	Brasil	América Central	México
Soya	15	30	—	5
Frutales	20	r	r	1
Algodón	15	30	82	50
Maíz/sorgo	10	—	r	10
Trigo	10	—	—	r
Hortalizas	10	11	r	15
Tabaco	10	—	r	r
Banano	—	r	r	r
Caña de azúcar	—	r	r	r
Arroz	—	—	6	1

r: resto.
Fuente: Ref. [6].

CUADRO II. DISTRIBUCION DEL USO DE HERBICIDAS POR CULTIVO EN AMERICA LATINA

Cultivo	Uso (en %)			
	Argentina	Brasil	América Central	México
Maíz	5	10	r	24
Caña de azúcar	20	10	32	15
Arroz	20	25	25	15
Pastos	—	10	—	10
Algodón	<20	8	18	10
Trigo	20	—	—	10
Sorgo	—	—	r	10
Soya	20	30	—	—
Café	—	7	10	—
Bananos	—	—	14	—

r: resto.
Fuente: Ref. [6].

Después de que el Area Sur, constituida por Argentina, Brasil, Chile, Paraguay y Uruguay y reunida en Santiago en 1983, aceptara las mismas normas [12] se puede decir que se ha logrado la armonización de 26 países, en uniformidad de la etiqueta, en derechos de propiedad de datos y en procedimientos y requisitos de registro.

2.5. Uso de plaguicidas en la agricultura en América Latina

Arauz et al. [13] encontraron que el 87% de los agricultores utiliza únicamente el combate químico, un 12% usa combate químico y cultural y solamente un 1% no usa métodos de lucha contra las plagas.

Las recomendaciones de los extensionistas agrícolas de los ministerios de agricultura tienen por lo tanto su efecto. El Manual de Recomendaciones del Ministerio de Agricultura y Ganadería (MAG), como señala Hilje [14], recomienda

CUADRO III. DISTRIBUCION DEL USO DE FUNGICIDAS POR CULTIVO EN AMERICA LATINA

Cultivo	Uso (en %)			
	Argentina	Brasil	América Central	México
Frutales	20	11	r	5
Papas	20	r	5	5
Cítricos	10	r	r	r
Viñedos	10	r	—	r
Hortalizas	10	30	5	30
Trigo	r	26	—	20
Algodón	—	—	—	5
Café	—	15	10	r
Arroz	r	5	5	r
Cacao	—	5	r	r
Bananos	—	—	62	r
Tabaco	—	—	5	r

r: resto:.
Fuente: Ref. [6].

en un 96% de los casos el uso de los plaguicidas y solamente en un 4% se recomiendan métodos alternativos.

En Argentina [6], más del 20% de insecticidas se utiliza en frutales, cerca del 15% en soya, 15% en algodón y algo menos del 10% en cada una de las siguientes cosechas: maíz, sorgo, trigo, hortalizas y tabaco. El uso de insecticidas en Brasil se distribuye en un 30% en soya, otro tanto en algodón, un 11% en hortalizas, y el resto en frutales, bananos y caña de azúcar. En el Cuadro I podemos comparar estos valores con los de América Central y México.

En el Cuadro II se puede comparar la distribución del uso de herbicidas por cosecha para Argentina, Brasil, América Central y México.

En el Cuadro III se puede observar la distribución del uso de los fungicidas para esas regiones latinoamericanas.

Podemos concluir que tanto el algodón como la soya y el arroz son los cultivos que más requieren de insecticidas en América Latina, mientras que el arroz, la caña de azúcar y la soya son los cultivos que demandan un mayor volumen de herbicidas. Sin embargo, también aumenta el uso de herbicidas en los pastizales en América Latina.

En cuanto al uso de fungicidas, el cultivo del banano, así como el de las hortalizas, presenta las mayores demandas, como también sucede con el cultivo del café y el cultivo de los frutales, estos últimos del cono sur.

CUADRO IV. PRINCIPALES PLAGUICIDAS UTILIZADOS EN EL CULTIVO DEL CAFE

Nombre genérico	Acción biocida	Clasificación química
Carbofuran	Nematicida	Carbamato
Etoprop	Nematicida	Organofosforado
Fenamifos	Nematicida	Organofosforado
Aldicarb	Nematicida	Carbamato
Terbufos	Nematicida	Organofosforado
Paraquat	Herbicida	Bipiridilo
2,4-D	Herbicida	Fenoxiacético
Misma	Herbicida	Arsenical
Glifosato	Herbicida	Aminoácido
Oxifluorfen	Herbicida	Difenileter
Terbutilazina	Herbicida	Triazina
Pendimetalin	Herbicida	Anilida
Dalapon	Herbicida	Acido alifático
Oxido de cobre	Fungicida	Cúprico
Oxicloruro de cobre	Fungicida	Cúprico
Captafol	Fungicida	Ditiocarbamato
Arseniato de plomo	Fungicida	Arsenical
Ferbam	Fungicida	Ditiocarbamato
PCNB	Insecticida	Organoclorado
Foxin	Insecticida	Organofosforado

Fuente: Ref. [3].

En el Cuadro IV, de manera ilustrativa se pueden observar los principales plaguicidas utilizados en Costa Rica en el cultivo de café, ya que es imposible entrar a describir aquí cada una de las cosechas importantes.

En América Latina podemos señalar tres categorías de usuarios de plaguicidas agrícolas:

a) Empresas transnacionales, que manejan cultivos extensivos para la exportación. Utilizan la tecnología agrícola más avanzada y criterios de uso de plaguicidas más estrictos. Resguardan su posición en el mercado internacional mediante un control de residuos y de calidad estrictos.

b) Grandes plantaciones nacionales, con cultivos también orientados hacia la exportación, y en menor grado al consumo interno. Tienen menos acceso al financiamiento y al uso de tecnologías más avanzadas. Ejercen un control cada vez más estricto sobre la aplicación de plaguicidas, para evitar problemas de residuos en sus productos.

c) Pequeños y medianos agricultores. Poseen grandes deficiencias de información, capacitación y financiamiento. Usan los plaguicidas con base en criterios muy cuestionables y son generalmente la fuente de problemas de contaminación y desarrollo de resistencias entre las plagas.

2.6. Uso de plaguicidas en salud pública

Los plaguicidas son también utilizados en el control de artrópodos y roedores, que son vectores de algunas enfermedades importantes como la malaria, la fiebre amarilla, la enfermedad de Chagas, la peste negra, el tifus, el dengue, filiariosis y otras.

En América Latina y el Caribe aproximadamente unos 231 millones de personas habitan áreas que originalmente eran maláricas. En la actualidad, gracias al uso de los plaguicidas, esas áreas se han reducido, por lo que el riesgo de contraer malaria afecta tan sólo al 25% de aquellas personas. Nadie puede negar que los plaguicidas han sido y siguen siendo un instrumento importante para reducir las enfermedades transmitidas por los insectos [15].

Lamentablemente, desde 1970 el número de casos de malaria ha aumentado, debido en parte al desarollo de la resistencia de los vectores a los plaguicidas utilizados en la agricultura, especialmente en la producción de algodón en América Central.

Actualmente más de 400 insectos han desarrollado resistencia a uno o más tipos de insecticidas [16].

En 1976, la OMS [17] informó de un aumento en la resistencia de las plagas de artrópodos a los plaguicidas. Entre los mosquitos, *Anopheles* spp., hacia 1969, 15 especies habían desarrollado resistencia al DDT. La resistencia al dieldrín se desarrolló con mayor celeridad aún: 37 especies en 1969. En 1976 aumentaron a 43.

CUADRO V. USO DE PLAGUICIDAS EN SILVICULTURA

Nombre genérico	Acción biocida	Clasificación química
Mirex	Insecticida	Organoclorado
Aldrín	Insecticida	Organoclorado
Metilparatión	Insecticida	Organofosforado
Foxín	Insecticida	Organofosforado
Malatión	Insecticida	Organofosforado
Acefato	Insecticida	Organofosforado
Metomil	Insecticida	Carbamato
Captafol	Fungicida	Dicarboximida
Oxicloruro de cobre	Fungicida	Inorgánico
Sulfato de cobre	Fungicida	Inorgánico
Paraquat	Herbicida	Bipiridilo
2,4-D	Herbicida	Fenoxiacético
Oxifluorfen	Herbicida	Difenileter
Glifosato	Herbicida	Aminoácido

Fuente: Ref. [19].

Otras 24 especies eran también resistentes al DDT, cinco a los organofosforados y dos a los carbamatos.

En los mosquitos del género *Culex,* que incluyen los vectores de fiebre amarilla, filariasis y el dengue, la resistencia había aumentado de 19 especies en 1968 a 41 en 1975, registrándose nuevamente varios casos de resistencia múltiple.

Las cantidades de plaguicidas destinadas al control de vectores son relativamente bajas y la responsabilidad de estas campañas recae generalmente en los ministerios de salud de los países latinoamericanos. Así, en Costa Rica, en 1983 se utilizó un 0,06% del total de plaguicidas para el combate de los mosquitos anofelinos. Se utilizó DDT y Baygon para un total de 6870 kg, de los cuales el 91% correspondió al DDT. En Guatemala se utilizó en 1984 únicamente un 2% del volumen total de plaguicidas en el combate de los vectores, mientras que en 1968 ese porcentaje fue del 6%.

La exposición de los habitantes de las zonas rociadas es alta, si consideramos que se rocían sus casas de habitación tanto externa como internamente. Umaña y Constenla [18] hallaron niveles más altos de DDT y sus metabolitos en leche materna en zonas donde la aplicación de DDT en la lucha antimalárica ha sido mayor.

2.7. Uso de los plaguicidas en la ganadería

La ganadería es una actividad de gran importancia en América Latina y requiere de plaguicidas para el combate de ectoparásitos y de moscas. Las cantidades de insecticidas utilizadas para este fin son relativamente pequeñas en comparación con las usadas en la agricultura.

Los productos más utilizados contra artrópodos como la garrapata son: el coumafos (Asuntol), tricrorfon (Neguvón), diclorvos (Nuvan), y dicritofos (Ektafos), todos los cuales son organofosforados.

2.8. Uso de plaguicidas en silvicultura

Hilje et al. [19] reportan sobre el uso de plaguicidas en silvicultura.

La silvicultura es una actividad nueva para América Latina. En el Cuadro V se recopilan los datos sobre el uso de plaguicidas en esta actividad en Costa Rica. La lista incluye productos muy tóxicos a corto plazo, como el metilparatión, el metomil y el paraquat, además de otros como el 2,4-D, el mirex, el aldrín y el captafol.

3. CONSECUENCIAS O IMPLICACIONES DEL USO DE PLAGUICIDAS EN AMERICA LATINA

El valor de la utilización de plaguicidas para las economías latinoamericanas es inapreciable ya que son preponderantemente agrícolas y ganaderas. Esta herramienta indispensable no puede considerarse independiente del paquete tecnológico que incluye la mecanización, fertilización, irrigación, manejo de suelos y mejoramiento genérico. En la mayoría de los casos el nivel de empleo racional de plaguicidas es inadecuado a los requerimientos necesarios para producir alimentos suficientes.

Los plaguicidas se aplican de muchas maneras: en la curación de semillas, en baños de inmersión, en aplicaciones localizadas, en inyecciones, en tratamiento de rociado aéreo y en tratamientos de rociado foliar. Cada uno de estos métodos trae consigo riesgos ambientales. Por ejemplo, en el curado de las semillas se utilizan cantidades muy pequeñas de productos químicos exactamente donde se necesitan, por lo que este uso no transmite residuos a las cosechas. En cambio, en el caso de las aplicaciones aéreas se aplican grandes cantidades de plaguicidas sobre grandes áreas, lo que potencialmente genera un riesgo ambiental muy considerable.

3.1. Consecuencias agroecológicas

La utilización unilateral, desmedida e irracional de plaguicidas puede tener efectos agroecológicos inconvenientes e indeseables [20].

En los monocultivos, la sobreabundancia, la concentración física y la uniformidad genética y cronológica del cultivo resultan ser los factores primarios que justifican la aparición de las plagas.

El factor secundario es el mecanismo por el cual los organismos que se encuentran en un equilibrio natural alcanzan el estado de plaga. Según Huffaker [20], el mecanismo es complejo y abarca las siguientes cuestiones:

a) Con el uso de los plaguicidas, los enemigos naturales de las plagas, como depredadores, parasitoides y patógenos, son afectados adversamente. En general, se asume que los enemigos naturales de las plagas son más susceptibles a los plaguicidas de amplio espectro de acción que los insectos herbívoros.

b) El mal empleo de los plaguicidas puede conducir a la "creación" de plagas. Generalmente se trata de un cambio de estado de plaga secundaria a primaria. Stephens [21] ilustra el efecto de los plaguicidas sobre los enemigos naturales de las plagas secundarias y la subsecuente "creación" de plagas primarias. Antes de 1950 había en Costa Rica dos plagas importantes en el cultivo del banano, una de ellas exótica (*Cosmopolites sorditus*). A mediados de la década de los cincuenta aparecieron dos plagas adicionales, *Castniomera humboldti* y *Platynota rostiana*, a pesar de la aspersión masiva de dieldrín. Al final de la década habían surgido siete especies más como plagas primarias, una de ellas *Antichloris viridis*, resistente al dieldrín. En una década se pasó de dos a once plagas, a causa del efecto adverso del dieldrín, directo o indirecto, sobre las poblaciones de enemigos naturales.

c) Huffaker [20] señala que ciertos plaguicidas afectan la fisiología de las plantas y los procesos bioquímicos y que en algunos casos esto tiene como consecuencia un aumento en la fecundidad de las plagas, por ejemplo, los ácaros.

d) El desarrollo de la resistencia es un fenómeno extremadamente dinámico. Georghiou [22] ha logrado identificar 414 casos de resistencia en 1980. Para complicar el cuadro, existe no solamente resistencia singular, sino también resistencia múltiple y resistencia cruzada. Un ejemplo de América Central con graves consecuencias para la salud es la resistencia del mosquito *Anopheles albimanus*. En efecto, resultó resistente a varios productos organofosforados (metilparatión, malatión y fenitrotión) utilizados en la agricultura, y al ser expuesto en laboratorio a los carbamatos (propoxur y carbaril) desarrolló resistencia a ellos.

e) Los agentes polinizadores bióticos entre los insectos son los lepidópteros y los hemnópteros. En América Latina parte de la producción agrícola depende de la participación de estos insectos. La muerte masiva por exposición directa o indirecta de abejas ha sido documentada [3].

3.2. Consecuencias económicas

Según el informe ROCAP/USAID [23], las plagas y las enfermedades causan pérdidas equivalente al 25-40% del potencial total de la producción de América Central. Una proyección de pérdidas en el mismo orden puede señalarse para el resto de América Latina.

Sin los plaguicidas las pérdidas serían mayores que un 40% del potencial de la producción. Debemos, sin embargo, entrar en un proceso que equilibre el balance de los beneficios con los costos directos e indirectos.

3.2.1. Costos directos

Los costos directos de los insumos son muy elevados.

El efecto económico de las plagas y su control en América Central ha sido estimado entre dóls. 650 y 800 millones anualmente, sin incluir los costos indirectos [23].

CUADRO VI. PERDIDAS ESTIMADAS DE GRANOS BASICOS OCASIONADAS POR PLAGAS EN AMERICA CENTRAL[a], PARA 1987 (en 1000 t)

Cultivo	Producción potencial	Producción real	Diferencia	Pérdidas post-cosecha	Total de pérdidas	Valor de pérdidas (millones de dóls.) (precios locales)
Arroz	509	393	117	28	195	28,3
Maíz	4376	3397	979	360	1339	104,4
Sorgo	462	342	120	57	171	26,5
Frijol	317	238	79	48	127	26,5
Total					1832	172

[a] Incluye: Guatemala, El Salvador, Costa Rica, Honduras y Nicaragua.
Fuente: Adaptado de la Ref. [23].

CUADRO VII. COSTOS DE APLICACION DE PLAGUICIDAS POR CADA CULTIVO EN GUATEMALA (1986)

Cultivo	dóls./ha
Arveja	312
Banano	308
Algodón	262
Tabaco	120
Café	107
Hule	92
Tomate	84
Melón	79
Caña	72
Cacao	67
Cereales	42
Frijol	36
Ajonjolí	20
Maní	13

Fuente: Adaptado de la Ref. [4].

3.2.2. Costos indirectos

Los costos indirectos en términos de efectos secundarios, ambientales, sociales y macroeconómicos deben reducirse.

El costo es una función de tres grupos de factores principales:

a) Pérdidas de producción debidas a plagas.
b) Inversiones de insumos para el control de plagas.
c) Costos macroeconómicos, sociales y ambientales, a menudo relacionados con el uso ineficaz o inadecuado de los plaguicidas. La importancia relativa de cada uno de estos factores varía según el cultivo, el sistema, la escala de empresa agrícola y los tipos de plagas.

CUADRO VIII. MORTALIDAD CAUSADA POR INTOXICACIONES AGUDAS EN COSTA RICA

Año	Organoclorados			Organofosforados y carbamatos			Otros		
	N° hospitalizaciones	N° defunciones	%	N° hospitalizaciones	N° defunciones	%	N° hospitalizaciones	N° defunciones	%
1976	21	1	5	—	—	—	191	16	8
1982	16	2	13	177	8	4,5	196	24	12
1983	13	0	0	275	11	4	214	41	19

Fuente: Ref. [19].

3.2.2.1. Pérdidas de producción

Como ejemplo, el informe ROCAP/USAID [23] estima las pérdidas en América Central en el año 1982 en cultivos de granos básicos en dóls. 172 millones, lo que representaba 1,8 millones de toneladas métricas (Cuadro VI).

Las pérdidas causadas por las plagas de los cultivos de exportación han sido muy elevadas durante muchos años y en algunos cultivos como el banano han asumido el carácter de crisis económica, ya que los elevados costos del combate, en este caso de la sigatoka negra, ha reducido la rentabilidad del cultivo. Esto ha tenido graves repercusiones en los niveles económicos y sociales.

3.2.2.2. Costos de control de plagas

Los plaguicidas representan el mayor porcentaje de los gastos para controlar las plagas de los cultivos. La proporción de la mano de obra necesaria para la aplicación es pequeña en América Latina. En el Cuadro VII se observan los costos de aplicación de plaguicidas por cultivo en Guatemala. El costo de los plaguicidas, incluyendo tanto los productos como la mano de obra involucrada, alcanzan el 30–35% del total de los costos del cultivo de bananos en Costa Rica [24]. Estos costos suben cada año, mientras que la eficacia del control de plagas, sobre todo el de la sigatoka negra, disminuye. En América Central los costos en plaguicidas en el cultivo del algodón corresponden al 40–50% de los insumos totales. Esta situación a contribuido mucho a la quiebra de muchas empresas agrícolas.

3.2.2.3. Costos macroeconómicos sociales y ambientales

Los costos indirectos del uso ineficaz o inadecuado de los plaguicidas se pueden evaluar teniendo en cuenta:

— El efecto sobre la balanza de pagos por razón de la importación.
— Los rechazos a las exportaciones de carnes y otros alimentos en los mercados de EE UU y Europa por la presencia de niveles no aceptables de residuos de plaguicidas.
— Los efectos ambientales y los efectos en la salud causados por el uso inadecuado de plaguicidas significan también costos económicos importantes.

Desde luego el costo de las intoxicaciones es difícil de cuantificar, sobre todo la pérdida de vidas humanas (Cuadro VIII).

Además de los riesgos y costos por intoxicaciones agudas se hace necesario cuantificar otros relacionados con la salud humana.

Los costos ambientales causados por el mal manejo de los plaguicidas incluyen la mortalidad masiva de aves, peces y otras faunas silvestres, y la contaminación de aguas y suelos, lo que puede afectar directamente la salud humana y animal.

En algunos casos, la sobreutilización de plaguicidas ha provocado efectos muy graves y permanentes en los suelos. Cordero y Ramírez [25] reportan la sobreacumulación de residuos de cobre en suelos dedicados, de 1930 a 1950, al cultivo del banano en el Pacífico Sur de Costa Rica. Las aplicaciones muy numerosas y concentraciones muy elevadas de caldo bordelés para combatir enfermedades fungosas causó la esterilización de los suelos en 50 000 ha, es decir causó la destrucción permanente de su capacidad de producción. Este caso representa una enorme pérdida económica en términos productivos.

4. INVESTIGACION Y TENDENCIAS

4.1. El manejo integrado de plagas

Desde que Bartlett en 1956 introdujo el término de lucha integrada, éste tuvo buena acogida y fue adaptado rápidamente, sobre todo en los países desarrollados.

La definición más reciente de manejo integrado se debe a Smith y van den Bosch [26]: "Sistema de lucha de una población de insectos dañinos en la que, teniendo en consideración el medio y la dinámica de la especie, se emplean todas las técnicas y métodos idóneos, de la manera más compatible posible, y se mantiene la densidad a un nivel tan bajo que no pueda causar perjuicios económicos."

En América Latina ha surgido la noción de que, más que remediar problemas, es necesario evitarlos, lo que implica entender y manejar los procesos ecológicos que se dan en los ecosistemas naturales y en los agroecosistemas.

El el manejo integrado de las plagas, el uso de los plaguicidas no se elimina, sino que se racionaliza, de modo que se utilicen sólo cuando sean necesario y en dosis y áreas definidas, con base en criterios precisos. Para ello se desarrollan las siguientes áreas de investigación [19]:

1) Identificación de las plagas claves de los cultivos más importantes.
2) Bionomía y fenología de las plagas.
3) Investigación en métodos de muestreo para la determinación rápida, eficiente y confiable de las plagas en el campo. Este conocimiento permitirá detectar y evaluar la magnitud de los problemas de plagas y tomar decisiones atinadas para su control.
4) Determinación y establecimiento de umbrales económicos. Este conocimiento es fundamental para el control de una plaga determinada y para la aplicación de plaguicidas a intervalos regulares.
5) Investigación en prácticas agrícolas, para sistematizar prácticas desfavorables a las plagas.
6) Control biológico.
7) Mejoramiento genético.
8) Plaguicidas.

La investigación en plaguicidas se dirige a determinar las dosis a utilizar en las diferentes condiciones y a establecer métodos que permitan evitar la resistencia de las plagas. Armonizar el uso de plaguicidas con el ambiente y la salud.

La investigación realizada en la región latinoamericana es cuantiosa, pero se carece de un índice completo de las publicaciones científicas sobre plagas agrícolas. Tan sólo en Costa Rica, según el índice de Jirón y Sancho de Barquero [27] se conocían, hasta 1980, 1598 publicaciones sobre insectos, de las cuales 172 tienen relación con plagas.

4.2. Protección del ambiente y la salud

La investigación se orienta a desarrollar una metodología para detectar y para predecir los problemas de los plaguicidas en el ambiente, a corto y largo plazo. En esta investigación tiene importancia el análisis de residuos en diferentes sustratos; la comparación de resultados entre zonas agrícolas y no agrícolas; el estudio de los efectos adversos de los plaguicidas sobre la fauna; y las investigaciones de campo y de laboratorio para determinar el movimiento de los plaguicidas en los ecosistemas, los tiempos de persistencia en los diferentes sustratos, su metabolismo y degradación, y su posible acumulación y biomagnificación.

En salud, la investigación debe contemplar aspectos toxicológicos, salud laboral, salud pública y residuos en alimentos.

Se han hecho muchos esfuerzos para dotar a la agricultura de plaguicidas cada vez más eficientes y seguros, desde que en 1857 se descubriera el acetoarsenito de cobre, el primer insecticida sintético usado por el hombre. La tecnología por sí misma no es un sustituto del pensamiento, y tampoco podemos esperar que traiga progreso y prosperidad sin efectos colaterales [28]. En América Latina los esfuerzos tendrán que multiplicarse para garantizar un uso óptimo de la tecnología de los plaguicidas.

REFERENCIAS

[1] BUECHEL, K.H., Impact of the agrochemicals industry on the third world, Chem. Ind. (London) (1979) 791–795.

[2] WILLIAMS, C.M., La tercera generación de plaguicidas, Sci. Am. **217** (1967) 13–17.

[3] CASTILLO, L.E., WESSELING, C., "Diagnóstico de la problemática de plaguicidas en Costa Rica", Seminario sobre los Problemas Asociados con el Uso de Plaguicidas en Centroamérica y Panamá, Instituto Interamericano de Cooperación para la Agricultura, San José, Costa Rica (1987).

[4] DE CAMPOS, M., "Problemas asociados con el uso de plaguicidas en Guatemala", ibid.

[5] FOOD AND AGRICULTURE ORGANIZATION OF THE UNITED NATIONS, Impact Monitoring of Residues from the Use of Agricultural Pesticides in Developing Countries, FAO/UNEP Experts Report, FAO, Rome (1975).

[6] MALTBY, C., Use of Pesticides in Latin America, Rep. UNIDO/IOD.353, United Nations Industrial Development Organization, Vienna (1980).

[7] LLERA, H., Asistencia Preparatoria para Determinar Necesidades y Objetivos en Material de Cooperación entre Países para la Elaboración de Plaguicidas en América Latina, Inf. UNIDO/IOD.543, Organización de las Naciones Unidas para el Desarrollo Industrial, Viena (1983).

[8] AGUILAR, J., et al., Generación y Transferencia Tecnológica Privada en el Sector Agrícola de Costa Rica: el Caso de los Agroquímicos, Proyecto COS 81/Tor, Ministerio de Planificación, San José, Costa Rica (1983).

[9] ORGANISMO INTERNACIONAL REGIONAL DE SANIDAD AGROPECUARIA, Reunión para Unificar Criterios sobre Registro, Etiquetado y Control de Calidad de los Plaguicidas en Centro América y Panamá, Isla Contadora, Panamá, OIRSA, San Salvador, El Salvador (1981).

[10] INSTITUTO INTERAMERICANO DE COOPERACION PARA LA AGRICULTURA, Reunión de Consulta sobre la Armonización de Etiquetado y Registro de Plaguicidas para los Países del Area Andina, Cartagena, Colombia, IICA (1982).

[11] INSTITUTO INTERAMERICANO DE COOPERACION PARA LA AGRICULTURA/OIRSA, II Reunión de Consulta para la Armonización de Criterios en Registro y Etiquetado de Plaguicidas para los Países del Area Central, San José, Costa Rica, IICA (1983).

[12] INSTITUTO INTERAMERICANO DE COOPERACION PARA LA AGRICULTURA, I Reunión de Consulta para la Armonización de Criterios en Registro y Etiquetado de Plaguicidas para los Países del Area Sur, Santiago, Chile, IICA (1983).

[13] ARAUZ, L.F., CARAZO, E., MORA, D., Diagnóstico sobre el uso y manejo de plaguicidas en las fincas hortícolas del Valle Central de Costa Rica, Agron. Cienc. **13** (1983) 37–49.

[14] HILJE, L., Estado actual de combate de plagas agrícolas en Costa Rica, Cienc. Amb. (1984) 5–6.

[15] ORGANIZACION PANAMERICANA DE LA SALUD, Health Conditions in the Americas, 1977-1980, Scientific Publication No. 427, Washington, DC (1982).

[16] GEORGHIOU, G.P., "The surveillance of pest resistance to insecticides in agriculture and public health", Proc. Int. Workshop, Sri Lanka (1982).

[17] WORLD HEALTH ORGANIZATION, Resistance of Vectors and Reservoirs of Disease to Pesticides, Technical Report Series No. 585, WHO, Geneva (1976).

[18] UMAÑA, V., CONSTENLA, M.A., Determinación de plaguicidas organoclorados en leche materna en Costa Rica, Rev. Biol. Trop. **32** 2 (1984) 233–239.

[19] HILJE, L., et al., El uso de plaguicidas en Costa Rica, Heredia (1986) (mimeografiado).

[20] HUFFAKER, C.R., "The ecology of pesticide interference with insect populations", Agricultural Chemicals, Harmony or Discord for Food, People and the Environment (SWIFT, J.E., Ed.), University of California, Los Angeles (1971) 22–104.

[21] STEPHENS, C.S., Ecological upset and recuperation of natural control of insect pests in some Costa Rican banana plantations, Turrialba **34** 1 (1984) 101–105.

[22] GEORGHIOU, G.P., Insecticide resistance and prospects for its management, Res. Rev. **76** (1980) 131–145.

[23] REGIONAL ORGANIZATION FOR CENTRAL AMERICAN PEST CONTROL, Integrated Pest Management, Project No. 596-0110, United States Agency for International Development, Washington, DC (1984) 175 pp.

[24] ASOCIACION BANANERA NACIONAL, Ref. citada en la Ref.[19] por Hilje et al.

[25] CORDERO, A., RAMIREZ, G., Acumulamiento de cobre en los suelos del pacífico sur de Costa Rica y sus efectos detrimentales en la agricultura, Agron. Costarricense, **3** 1 (1979) 63–78.

[26] SMITH, R.F., VAN DEN BOSCH, R., "Integrated control", Pest Control, Academic Press, New York (1967).

[27] JIRON, L.F., SANCHO DE BARQUERO, M.E., Indice de publicaciones Entomológicas de Costa Rica, San José, Costa Rica, Consejo Nacional de Investigaciones Científicas y Tecnológicas/Organization of Tropical Studies (1983) 305 pp.

[28] RUCKELSHAUS, W.D., The Environmental Crisis, Our Work has Just Begun, paper presented at the National Press Club, Washington, DC, Jan. 1971.

DISTRIBUTION OF RESIDUES OF ^{14}C-ALDICARB APPLIED TO COTTON PLANTS IN GEZIRA, SUDAN*

G.A. EL ZORGANI, T.N. BAKHIET,
N. SHARAF ELDIN
Entomology Section,
Gezira Research Station,
Agricultural Research Corporation,
Wad Medani, Sudan

Abstract

DISTRIBUTION OF RESIDUES OF ^{14}C-ALDICARB APPLIED TO COTTON PLANTS IN GEZIRA, SUDAN.
  The uptake and distribution of ^{14}C-aldicarb as well as the granular formulation Temik 15G were studied in cotton plants. The maximum uptake was recorded 2 weeks after application in the case of aqueous aldicarb treatments, while the maximum uptake from the granular formulation was recorded after 4 weeks. The highest content of aldicarb derived residues was invariably found in the leaf tissue. Extensive metabolism into highly polar compounds was evidenced. No aldicarb, its sulphoxide or sulphone were detectable in the plant tissue or in the soil at harvest.

1. INTRODUCTION

The carbamate insecticide/nematicide aldicarb (2-methyl-2-(methylthio) propionaldehyde 0-(methylcarbamoyl)-oxime) is marketed in granular formulations containing 10 and 15% aldicarb as Temik 10G and Temik 15G, respectively. The latter formulation was successful in the control of the cotton white fly *Bemisia tabaci* (Gen.), the major cotton pest in Gezira, Sudan [1]. The first large scale application during the 1979-1980 season (at a rate of 17 lb/feddan[1], 10 weeks after planting) demonstrated its effectiveness in reducing the white fly population as well as increasing the cotton yield [2]. Therefore, the Temik treated area in the Gezira increased progressively, reaching a maximum of nearly 200×10^3 feddan during the 1984-1985 season [3].
  The fate of aldicarb in cotton plants and soil and its metabolism in cotton foliage and seed under field conditions in the United States of America have been

---

* Research carried out with the support of the IAEA under Technical Co-operation Project SUD/5/012.

[1] 1 lb = 0.4536 kg; 1 feddan = 0.42 ha.

reported [4, 5]. Aldicarb sulphoxide (2-methyl-2-(methylsulphinyl)propionaldehyde 0-(methylcarbamoyl)-oxime) and sulphone (2-methyl-2(methylsulphonyl)propionaldehyde 0-methylcarbamoyl)-oxime) were identified as the main metabolites in cotton. The structures of these substances are shown in Fig. 1. They are, however, unstable and are transformed within a short period (t½ sulphoxide: 1 week) into a number of highly polar derivatives.

Aldicarb is characterized by its high acute toxicity (rat oral LD_{50} = <2 mg/kg). The current experiments were undertaken to assess the magnitude of its toxic residues in cotton products and in soil under Gezira conditions. Another objective was to identify the dynamics and distribution of aldicarb in the crop in order to achieve better pest control through synchronization of the application date and the pest infestation peaks.

2. MATERIALS AND METHODS

2.1. Chemicals

Carbon-14-aldicarb of a specific activity (6.06 mCi/mmol)[2], aldicarb sulphoxide and sulphone were a gift from Union Carbide, USA. Temik 15G was also obtained from the same source. The radiochemical purity of ^{14}C-aldicarb was 99%, as determined by thin layer chromatography and a liquid scintillation counter. All the solvents used were of the reagent grade (Baker Analysed).

2.2. Uptake and distribution of ^{14}C-aldicarb by potted cotton plants

Two month old cotton plants (var. Barakat) were used. Three plants per pot were grown in 1.5 kg of heavy clay Gezira soil. Carbon-14-aldicarb of a specific activity $(1.46 \times 10^6 \text{ dis} \cdot \text{min}^{-1} \cdot \text{mg}^{-1})$[3] and at the rate of 30 mg per pot was applied. Four evenly spaced holes, 0.5 to 1 cm deep, were made on the soil surface of each pot and the prescribed amount of aqueous acetone solution of aldicarb was evenly distributed in the holes, giving a soil aldicarb concentration of 20 ppm and a total radioactivity of 43.8×10^6 dis/min per pot. Twenty millilitres of liquid fertilizer solution (Wuxal suspension) were added to each pot. The treated pots were placed outdoors in a screened cage and watered twice daily. The plants were removed at intervals of 2, 4 and 6 weeks, extracted and analysed for radioactivity in the leaf, stem and roots. The plants were macerated in an acetone–water mixture (85:15), filtered, and the aliquots decolorized by exposure to sunlight. After the

---

[2] 1 Ci = 3.70×10^{10} Bq.

[3] 1 dis/s = 60 dis/min = 1 Bq.

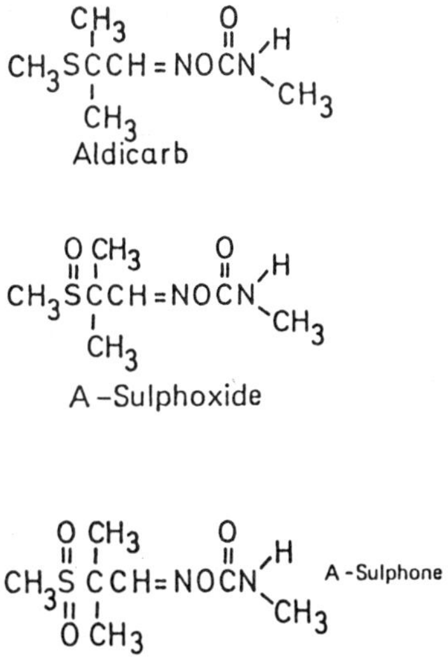

$$CH_3S\underset{\underset{CH_3}{|}}{\overset{\overset{CH_3}{|}}{C}}CH=NOC\overset{O}{\overset{\|}{N}}\underset{CH_3}{\overset{H}{\diagup}}$$

Aldicarb

$$CH_3\overset{O}{\overset{\|}{S}}\underset{\underset{CH_3}{|}}{\overset{\overset{CH_3}{|}}{C}}CH=NOC\overset{O}{\overset{\|}{N}}\underset{CH_3}{\overset{H}{\diagup}}$$

A-Sulphoxide

$$CH_3\underset{\underset{O}{\|}}{\overset{\overset{O}{\|}}{S}}\underset{\underset{CH_3}{|}}{\overset{\overset{CH_3}{|}}{C}}CH=NOC\overset{O}{\overset{\|}{N}}\underset{CH_3}{\overset{H}{\diagup}} \quad \text{A-Sulphone}$$

FIG. 1. Aldicarb and its main metabolites.

addition of the scintillation cocktail, samples were counted in a Packard 300 CD instrument [6].

2.3. Uptake and distribution of ^{14}C-aldicarb in field grown cotton plants

In a 4 month old cotton field cultivated according to Gezira standard practice, two plant holes, representing the two most common varieties Barakat and Acala, were selected. Each hole, containing three plants, was treated with 75 mg of ^{14}C-aldicarb of the same specific activity as that mentioned in Section 2.2. Treatment was applied immediately before the scheduled watering. After 6 weeks, the plants were uprooted and the radioactivity in various parts was determined. Ten weeks later the soil from each hole was sampled and analysed.

2.4. Uptake of Temik 15G by field grown cotton plants, var. Barakat

Eight week old field cotton plants (var. Barakat) were treated with Temik 15G at the rate of 17 lb/feddan using the side dressing method, and were then immediately watered. The treated plots formed part of an application date trial in a completely randomized design replicated six times. Three plants were removed from each replicate at 2, 4 and 6 week intervals after application. Leaves were excised, mixed and subsampled for extraction and analysis by gas–liquid chromatography (GLC), as described in the following section.

TABLE I. UPTAKE AND DISTRIBUTION OF ^{14}C-ALDICARB IN POTTED COTTON PLANTS

Time after application (weeks)	Total radioactivity extracted × 10^6 dis/min	Tissue distribution of radioactivity (% of total)		
		Leaf	Stem	Root
2	2.96 ± 0.21	76 ± 4	2.0 ± 0.2	22 ± 3
4	1.87 ± 0.18	77 ± 4	12.5 ± 1.5	10.5 ± 1.6
6	1.75	46.0	19.0	35

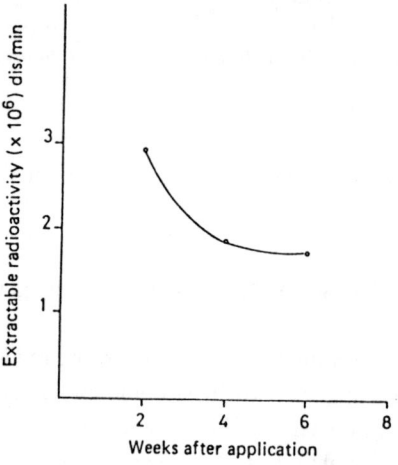

FIG. 2. *Uptake of ^{14}C-aldicarb in potted cotton plants.*

2.5. Analytical procedures

GLC analysis was performed by a method based on that described by Carey et al. [7]. Leaf samples were macerated in an acetone–dichloromethane (1:1) mixture in the presence of an equal weight of anhydrous sodium sulphate. The concentrated extract was then chromatographed on a Florisil column using hexane–acetone mixtures. Aldicarb and its sulphoxide were eluted by hexane–acetone (7:3) (F_1), while the sulphone containing fraction was eluted with hexane–acetone (1:1) (F_2). The first fraction was subjected to peracetic acid treatment before injection into the GLC

TABLE II. DISTRIBUTION OF EXTRACTABLE RESIDUES OF ^{14}C-ALDICARB IN FIELD COTTON PLANTS 6 WEEKS AFTER APPLICATION

Variety	Radioactivity × 10^3 dis/min						
	Leaf	Stem	Boll green	Dry boll shell	Seed	Root	Total × 10^3 dis/min
Var. Barakat							
Plant 1	974.4 (50.0)	133.4 (6.9)	146.1 (7.5)	440.1 (22.8)	186.2 (9.6)	50.8 (2.6)	1.931
2	2413.2 (74.4)	56.2 (1.7)	242.6 (7.0)	257.2 (7.9)	175.0 (5.3)	93.3 (2.8)	3.241
3	121.9 (10.9)	70.0 (6.3)	373.9 (33.7)	338.2 (30.5)	153.0 (13.8)	50.7 (4.5)	1.107
Var. Acala							
Plant 1	1319.0 (47.8)	293.0 (10.6)	280.0 (10.1)	51.8 (1.8)	600.0 (21.7)	214.0 (7.7)	2.757
2	1134.0 (48.7)	207.0 (8.8)	233.0 (10.0)	51.2 (2.1)	93.6 (3.9)	608.5 (26.1)	2.327
3	603.2 (24.6)	293.0 (11.9)	—	227.4 (9.2)	310.8 (12.6)	1011.8 (41.3)	2.446

Note: Numerals in brackets denote percentage.

instrument; F_2 was injected directly. Analysis was carried out on a Carlo Erba gas chromatograph fitted with a Perkin–Elmer flame photometric detector operated with an S filter.

3. RESULTS AND DISCUSSION

The pattern of uptake and distribution of ^{14}C-aldicarb in potted cotton plants is shown in Table I and represented graphically in Fig. 2. The maximum uptake was recorded at the first sampling, 2 weeks after application. At all sampling intervals leaf tissue had the highest content of radioactivity.

Six weeks after application of ^{14}C-aldicarb to field cotton plants, considerable amounts of radioactivity were found in all the components. The relative distribution of radioactivity is shown in Table II and a comparison between the distribution in

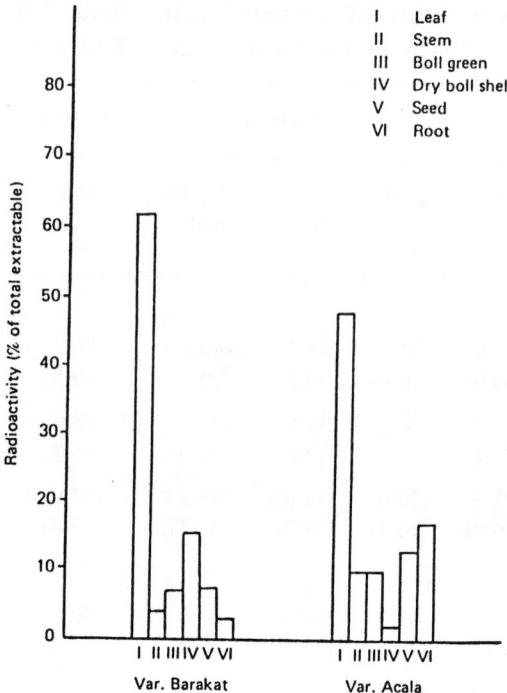

FIG. 3. *Distribution of ^{14}C-aldicarb residues in field cotton plants 6 weeks after application.*

the two cotton varieties is given in Fig. 3. Aqueous acetone extracts of the treated plants were partitioned in dichloromethane, after removal of the acetone. In all cases, more than 50% of the radioactivity remained in the aqueous phase, which is consistent with the recognized extensive metabolism of aldicarb into highly polar derivatives. The hydrophilic portion of radioactivity was particularly high in green boll and seed extracts. The soil contained no measurable radioactivity 16 weeks after application.

Both experiments with labelled aldicarb have demonstrated the importance of leaves as the main site of accumulation for aldicarb and metabolites. Pronounced variability between Barakat replicates (Table II) is due to loss of leaves. Leaf analysis from Temik 15G treated plants showed a maximum uptake 4 weeks after application, measured as aldicarb, aldicarb sulphoxide and aldicarb sulphone. The average total residue levels were 5.3, 12.65 and 7.13 ppm at 2, 4 and 6 weeks, respectively, after application (see Fig. 4). At harvest, the soil, foliage and seed from treated plants contained no residues.

At all sampling intervals, the ratio of the residue content between F_1 and F_2 was about 4:1. This is understandable in view of the fact that sulphoxide is not solely

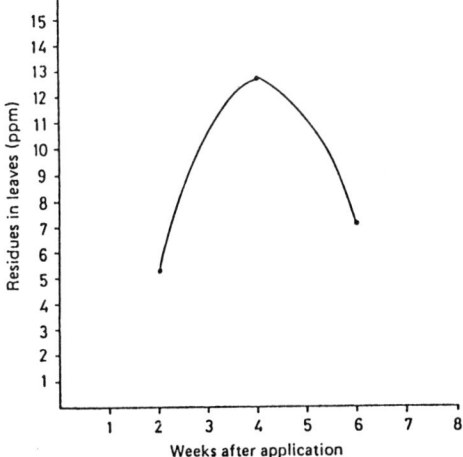

FIG. 4. *Uptake of Temik 15G in field cotton plants (var. Barakat).*

a precursor for sulphone, but is rather degraded primarily by hydrolysis to a sulphinyl oxime [4], which cannot be detected by the method employed.

On the basis of the above results we can safely conclude that aldicarb, when applied to Gezira cotton for up to 8 weeks after planting, leaves no known toxic residues in the plant products or in the soil at harvest. The maximum uptake, and consequently the best control of sucking pests, should be expected about 4 weeks after application.

ACKNOWLEDGEMENTS

The authors are grateful to Mark Deng Awin for technical assistance and to the IAEA for financial support.

REFERENCES

[1] BALLA, A.N., Recommendation of Temik 15G, Bidrin 103 + Endosulfan 50, Bidrin 24 + Endosulfan 50, Rogor 50 and Rogor 50 + DDT, Agricultural Research Corporation Report, Wad Medani, Sudan (1980).

[2] ABDUL-FARRAG, A., Results of Application of Temik 15G during the Seasons 1979/80 and 1980/81, Sudan Gezira Board Report (1981) (in Arabic).

[3] SUDAN GEZIRA BOARD, Annual Agricultural Report, 1985.

[4] COPPEDGE, J.R., LINDQUIST, D.A., BULL, D.L., DOROUGH, H.W., Fate of 2-methyl-2-(methylthio)propionaldehyde 0-(methylcarbamoyl)-oxime (Temik) in cotton plants and soil, J. Agric. Food Chem. **15** 5 (1967) 902.

[5] ANDRAWES, N.R., ROMINE, R.R., BAGLEY, W.P., Metabolism and residues of Temik aldicarb pesticide in cotton foliage and seed under field conditions, J. Agric. Food Chem. **21** 3 (1973) 379.

[6] EL ZORGANI, G.A., SHIMABUKURU, R., ABDULLA, A.M., AWIN, M.D., Technical Assistance Project Report, 1983, IAEA, Vienna (Internal report).

[7] CAREY, W.F., HELRICH, K., Improved method for the determination of aldicarb and its oxidation products in plant materials, J. Assoc. Off. Anal. Chem. **53** 6 (1970) 1290.

IAEA-SM-297/23

FATE AND MAGNITUDE OF MALATHION RESIDUES IN STORED WHEAT AND BARLEY*

K. GÖZEK, F. ARTIRAN
Department of Chemistry,
Ankara Nuclear Research and
  Training Centre,
Turkish Atomic Energy Authority,
Ankara, Turkey

Abstract

FATE AND MAGNITUDE OF MALATHION RESIDUES IN STORED WHEAT AND BARLEY.
  Wheat and barley, treated with ^{14}C-malathion, were stored in two different types of Central Anatolian conditions for 9 months. The sampling times were 0, 1/2, 1, 2, 3, 6 and 9 months. The surface, extractable and bound residues and the effect of baking were investigated. Surface and extractable ^{14}C-malathion residues were affected by storage time, but not by the type of storage. There is no significant difference between residues in grains stored in wooden boxes and those stored in enamelware buckets. The results show that surface and extractable residues increase with storage time. Bound residues were negligible. The maximum value of malathion residues in the grain was 6.5 mg/kg.

1. INTRODUCTION

Malathion (0,0-dimethyl S-(1,2-dicarbethoxyethyl)phosphorodithioate) is an important and widely used pesticide. It has been accepted for the control of pests on vegetables and field crops, fruits and nuts, stored grains and domestic animals. Because of its low mammalian toxicity, the insecticide can be mixed directly with the grain, where it controls pests by both contact and vapour activity [1]. Malathion has been used as the main insecticide in stored grains in Turkey and has been applied at a rate of 500 g of 2% malathion dust per tonne of grain, or equal to 10 ppm of active ingredient.

This research was carried out to determine the malathion residues in wheat and barley under Central Anatolian conditions. The local type of storage using enamelware buckets or wooden boxes was simulated.

---

* Research carried out with the support of the IAEA under Research Contract No. 3476/RB.

2. MATERIALS AND METHODS

2.1. Chemicals

Carbon-14-malathion (prepared by the condensation of diethyl(3-^{14}C)maleate with 0.0-dimethyl dithiophosphoric acid) was purchased from Amersham, United Kingdom, and had a specific activity of 37 mCi/mmol.[1] Non-radioactive malathion was obtained from the Plant Protection Institute, Ankara. Methanol, toluene, dioxane, methylcellosolve, ethanolamine, fuming H_2SO_4 (20 to 30% SO_3), H_3PO_4 (85%), $K_2Cr_2O_7$, KIO_3 and stannous chloride were of analytical reagent grade. A scintillation cocktail was prepared by dissolving 7 g of PPO, 0.05 g of POPOP and 50 g of naphthalene in 1 L of dioxane [2].

2.2. Grain treatment

The 1984 wheat (Kıraç 66) and barley (Tokak) crops were obtained from a farm in a suburb of Ankara. They had not been treated with insecticides prior to use. Foreign materials and cracked grains were removed by hand [3]. Wheat or barley (1.5 kg) of each storage type were treated with 140 μCi of ^{14}C-malathion and 0.75 g of 2% non-radioactive malathion dust, equivalent to 10 ppm of active ingredient. The moisture contents of the wheat and barley were 9.2% and 6.2%, respectively. At these levels the growth of pests and other microorganisms is almost completely suppressed. The grains were stored at ambient temperature (10 to 30°C).

It was noted that the dust, although of commercial quality, did not adhere very well to the grain surface. Hence, it was difficult to obtain uniform treatments as some of the dust quickly settled to the bottom of the container.

2.3. Sampling

Zero-time samples were taken on 19 March 1985, 21.5 hours after treatment. Further samples were collected after 15 days and 1, 2, 3, 6 and 9 months. Forty gram grain samples were collected and triplicate analyses were carried out for each sampling time.

2.4. Residue determination

Determination of malathion residues was performed as described in Fig. 1.

---

[1] 1 Ci = 3.70 × 10^{10} Bq.

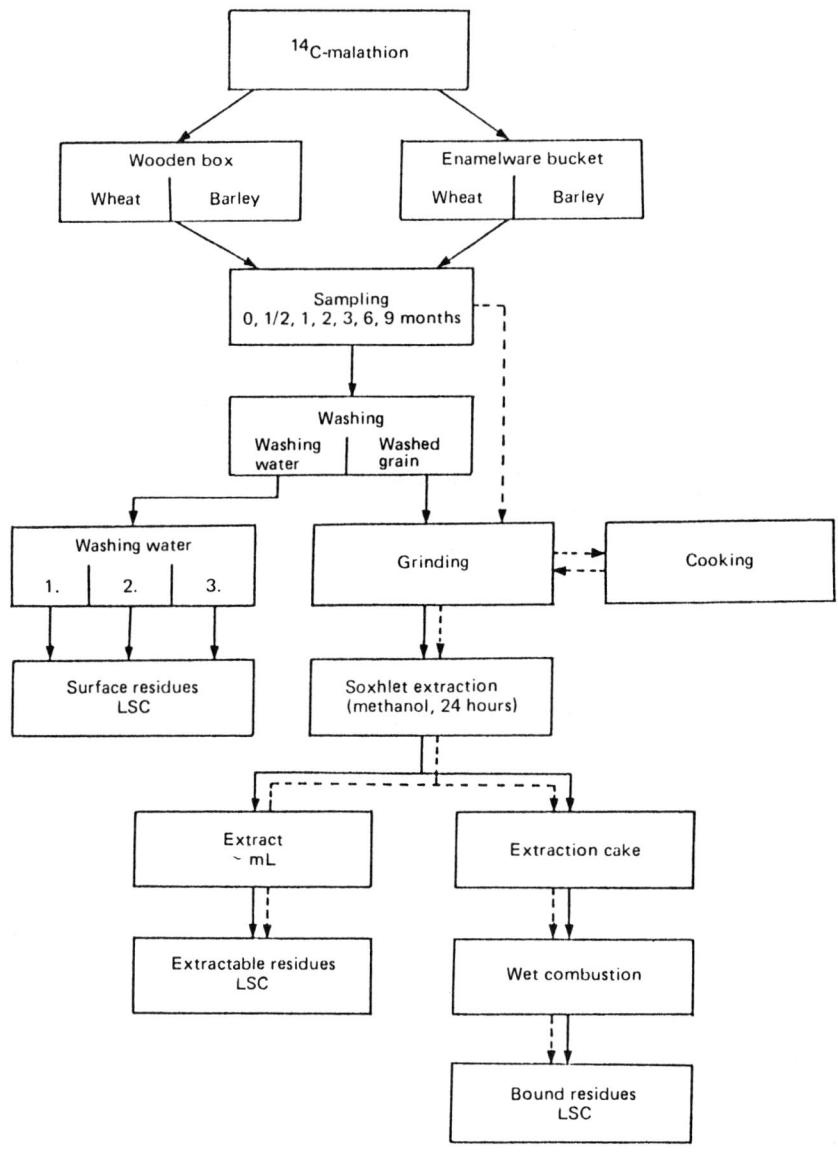

FIG. 1. Schematic diagram of residue analysis.

2.4.1. Surface residues

The grains were washed three times with 50 mL of distilled water to remove the surface residues. The radioactivity was measured for each separate washing.

2.4.2. Extractable residues

The washed grains were ground thoroughly and transferred to the thimbles of the Soxhlet apparatus. Analytical grade methanol was added and the grains were extracted for 24 h. After completion of the extraction, the volume of the extract was measured and an aliquot (2 mL) was counted by a liquid scintillator counter.

2.4.3. Non-extractable (bound) residues

(1) *Apparatus.* The wet combustion apparatus of Smith et al. [4] was constructed locally according to the original design; it consisted of a carbon dioxide scrubber, a combustion unit, an iodine trap and a carbon dioxide absorber.

(2) *Reagents.* These included: an ethanolamine solution: 70% of methylcellosolve mixed with 30% vol./vol. of ethanolamine; a scintillation solution: 8.25 g of PPO dissolved in 1 L of toluene; dry digestion reagent: KIO_3-$K_2Cr_2O_7$ (2:1, wt/wt); liquid digestion reagent: 1 g of KIO_3 dissolved in a mixture of 67 mL of fuming H_2SO_4 (20 to 30% SO_3) and 33 mL of 85% H_3PO_4 at a temperature of 160 to 190°C.

Analyses were carried out according to the procedure of Smith et al. [4]. For this purpose, the extraction cake was dried and 1.0 g portions were wet combusted; the samples were counted by LSC.

2.4.4. Effect of baking on the residue level

The baking process was performed after the last sampling. Doughs (100 g grain) were prepared in aliminium foil, by adding 110 mL of distilled water and baking at 200°C for 1 h.

For determination of the extractable residues, 40 g of bread samples were ground and extracted with methanol, as previously described. To determine the bound residues, the extracted bread was dried and 1.0 g portions were wet combusted and counted by LSC.

3. RESULTS AND DISCUSSION

3.1. Surface residues

Surface residues could not be completely removed by three washings and it is possible that a portion of the surface residue may have been included in the determination of the extractable residue. Tables I–IV show the surface residues for barley and wheat stored in wooden boxes and in enamelware buckets. These ranged from

TABLE I. DISTRIBUTION OF MALATHION RESIDUES IN BARLEY STORED IN WOODEN BOXES (in ppm active ingredient)

Storage time (months)	Surface residues (S)	Extractable residues (E)	Bound residues (B)	(S + E + B)
0	0.7	0.5	—	1.2
	0.6	0.5		1.1
	0.9	0.5		1.4
1/2	0.5	1.2	—	1.7
	0.9	1.1		2.0
	0.8	1.2		2.0
1	0.7	1.3	—	2.0
	0.8	1.3		2.1
	0.8	1.3		2.1
2	1.0	2.0	—	3.0
	1.1	1.8		2.9
	1.1	2.0		3.1
3	1.2	2.7	—	3.9
	1.4	2.7		4.1
	1.3	2.7		4.0
6	1.3	4.3	0.003	5.6
	1.2	4.3	nd	5.5
	1.1	3.4	nd	4.5
9	1.4	3.9	0.13	5.5
	1.4	3.7	0.1	5.2
	1.4	3.6	0.04	5.0

10 ppm = 100% of the theoretically applied dose, or activity.
nd = not detectable.

TABLE II. DISTRIBUTION OF MALATHION RESIDUES IN BARLEY STORED IN ENAMELWARE BUCKETS (in ppm active ingredient)

Storage time (months)	Surface residues (S)	Extractable residues (E)	Bound residues (B)	(S + E + B)
0	1.1	0.6	—	1.7
	1.0	0.7		1.7
	0.9	0.7		1.6
1/2	0.9	1.3	—	2.2
	1.0	1.3		2.3
	0.9	1.2		2.1
1	0.8	1.6	—	2.4
	0.6	1.6		2.2
	0.8	1.2		2.0
2	1.0	2.2	—	3.2
	1.3	2.2		3.5
	1.1	2.4		3.5
3	1.0	2.3	—	3.3
	1.1	2.4		3.5
	0.9	2.4		3.3
6	1.0	3.1	nd	4.1
	0.8	3.5	nd	4.3
	0.8	3.0	nd	3.8
9	1.4	4.0	0.01	5.4
	1.3	3.7	0.01	5.0
	1.6	4.1	0.002	5.7

10 ppm = 100% of the theoretically applied dose, or activity.
nd = not detectable.

TABLE III. DISTRIBUTION OF MALATHION RESIDUES IN WHEAT STORED IN WOODEN BOXES (in ppm active ingredient)

Storage time (months)	Surface residues (S)	Extractable residues (E)	Bound residues (B)	(S + E + B)
0	0.7 0.8 0.8	0.5 0.6 0.5	— — —	1.2 1.4 1.3
1/2	0.8 0.9 0.6	0.6 0.8 0.7	— — —	1.4 1.7 1.3
1	0.9 0.9 0.9	1.0 0.9 0.8	— — —	1.9 1.8 1.7
2	1.1 1.1 1.1	1.3 1.4 1.3	— — —	2.4 2.5 2.4
3	1.8 1.4 1.5	1.7 1.5 1.8	— — —	3.5 2.9 3.3
6	1.4 1.3 1.1	3.0 2.9 2.6	0.05 0.04 nd	4.4 4.2 3.7
9	2.5 1.9 2.1	3.9 3.2 3.7	0.14 0.10 0.14	6.5 5.2 5.9

10 ppm = 100% of the theoretically applied dose, or activity.
nd = not detectable.

TABLE IV. DISTRIBUTION OF MALATHION RESIDUES IN WHEAT STORED IN ENAMELWARE BUCKETS (in ppm active ingredient)

Storage time (months)	Surface residues (S)	Extractable residues (E)	Bound residues (B)	(S + E + B)
0	1.0	0.7	—	1.7
	1.0	0.6		1.6
	1.2	0.7		1.9
1/2	1.4	1.1	—	2.5
	1.1	1.3		2.4
	1.4	1.1		2.5
1	0.8	1.3	—	2.1
	1.4	1.2		2.6
	0.7	1.3		2.0
2	1.8	1.9	—	3.7
	1.5	1.8		3.3
	1.6	2.0		3.6
3	1.8	2.1	—	3.9
	1.7	2.1		3.8
	1.8	2.2		4.0
6	1.4	3.9	0.03	5.4
	1.1	3.0	0.02	4.1
	1.0	3.1	nd	4.1
9	1.7	3.8	0.01	5.5
	1.7	3.6	0.08	5.3
	1.7	3.5	0.12	5.3

10 ppm = 100% of the theoretically applied dose, or activity.
nd = not detectable.

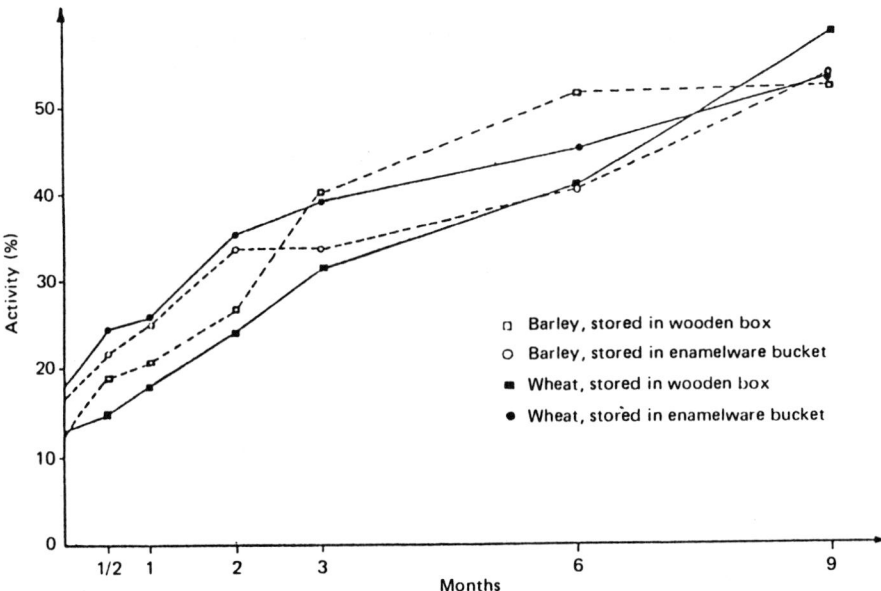

FIG. 2. Effect of storage time and type of storage for malathion residues on wheat and barley.

6 to 24% of the applied dose, with a tendency to increase with storage time. This is obviously due to the increased ability of the dust to adhere to the grain surface with time.

3.2. Extractable residues

Methanol extractable residues for barley stored in wooden boxes and in enamelware buckets increased from 5 to 41–43% (Tables I and II) during a storage period of 9 months. The corresponding values for wheat increased from 5–6 to 38–39%. These results show that the extractable residues and also the sum of surface and extractable residues increase with storage time to a maximum of 6.5 ppm for wheat after 9 months. The increase in extractable residues can only be caused by progressive diffusion into the grains. This is also illustrated in the tables, as the ratios between extractable and either surface or total residues are increasing with time. After 9 months the barley seems to have absorbed around three-quarters of the total content, whereas wheat absorbed two-thirds of the total content. It is probable that the malathion penetrates faster into barley than wheat. However, the value of 6.5 ppm is below the maximum residue limit of 8 ppm recommended by the Food and Agriculture Organization of the United Nations/World Health Organization. The results also show insignificant differences in the residue values obtained for grains stored in different types of containers (see Fig. 2).

TABLE V. EFFECT OF BAKING ON ^{14}C-MALATHION RESIDUES IN BARLEY AND WHEAT STORED FOR 9 MONTHS

Type of grain	Total residues[a] (S + E + B) (ppm)			
	Wooden box		Enamelware buckets	
	Before baking	After baking	Before baking	After baking
Barley	5.3	3.3	5.3	3.6
Wheat	5.9	1.9	5.4	2.7

[a] Data are the means of four replicates.

3.3. Bound residues

Up to 6 months the bound residues were mostly non-detectable or very low. At 9 months some samples gave a value of 0.14 ppm, which accounts for approximately 10% of the total grain residue (Table III).

3.4. Effect of baking

The total ^{14}C-malathion residues after the baking process are given in Table V. It can be seen that baking could lead to a substantial loss in malathion residues. In this connection, Acton and Parouchais [5] reported that considerable losses of malathion also occurred during the milling process.

4. CONCLUSIONS

After application of malathion dust, it has been demonstrated that it only partially adheres to the grain surfaces. The total malathion contents in the grain increase with time from only 10 to 20% of the applied dose to 50 to 60%. Simultaneously, the amount of malathion that is absorbed increases dramatically from only about 5% to nearly 40% of the applied dose. This is obviously caused by diffusion through the gas phase or solid diffusion. The increase in absorbed concentration also means that the longer the storage time, the less the possibility of removing the residues from the grain products through normal industrial processes such as washing, milling and fermentation. Baking may be one of the processes which can result in a decrease in the residue concentration in grain products such as bread, since around half the residues disappeared during the baking process; for wheat, the decrease was even greater.

REFERENCES

[1] ROWLANDS, D.G., HORLER, D.F., "Penetration of malathion into wheat grains", Proc. 4th Br. Insecticide Fungicide Conf., Brighton, 1967, Vol. 1, British Crop Protection Council, Thornton Heath, Surrey (1967) 331–335.

[2] L'ANNUNZIATA, M.F., Radiotracers in Agricultural Chemistry, Academic Press, New York and London (1979).

[3] ANDEREGG, B.N., MADISEN, L.J., Effect of dockage on the degradation of ^{14}C-malathion in stored wheat, J. Agric. Food Chem. **31** (1983) 700–704.

[4] SMITH, G.N., et al., Simple apparatus for combustion of samples containing ^{14}C-labelled pesticides for residue analysis, J.Agric. Food Chem. **12** 2 (1964) 172–175.

[5] ACTON, G.E., PAROUCHAIS, C., Food Technol. **18** (1966) 77.

IAEA-SM-297/31

EFFICACY OF THE CONTROLLED RELEASE OF ^{14}C-CARBOFURAN FORMULATION FOR PEST CONTROL IN COTTON*

F.F. JAMIL, M.J. QURESHI, A. HAQ,
N. BASHIR, S.H.M. NAQVI
Biological Chemistry Division,
Nuclear Institute for Agriculture and Biology,
Faisalabad, Pakistan

Abstract

EFFICACY OF THE CONTROLLED RELEASE OF ^{14}C-CARBOFURAN FORMULATION FOR PEST CONTROL IN COTTON.

Protocols for the treatment of cotton plants with a ring labelled ^{14}C-carbofuran formulation made up of EVA (polyethylene-polyvinyl acetate co-polymer), with a specific activity of 0.0169 mCi/g were modified in these studies as the dose per pot was reduced by half. Twenty-five plastic pots with a 30 cm diameter and lined with polyethylene film were filled with 22 kg of good homogenized soil. To 12 of these prepared pots, 0.5 g of formulation flakes (1.5 to 2 cm) containing 9.6 mg of carbofuran and with 8.45 µCi radioactivity were applied in a circle with a 5 cm diameter and at a depth of 5 cm. Four cotton seeds of the variety NIAB-78 were planted in the centre of each pot. Cotton seeds were planted in the remaining pots to produce the control plants. It has been observed that even the low amount of carbofuran formulation applied to the cotton crop (0.5 g/pot) offered some plant protection, but was not as effective as the higher dose (1 g/pot) applied in previous studies. Initially, the cotton plants looked healthier (up to 3 months) but then their resistance decreased in comparison to that of the control plants. It is suggested that the half dose applied in these studies is lower than desirable and that probably two-thirds of the dose could better serve the purpose. The radioactivity in the formulation flakes fell to almost one-half when the plants were 1 month old and then decreased gradually with the increase in time. In contrast, the radioactivity in the soil increased gradually with the increase in time. The highest radioactivity in the soil was observed from the 10 cm diameter samples at a depth of up to 13 cm. More radioactivity was recovered from the leaves than the roots and stem, which is in accordance with our previous findings.

---

* Research carried out with the support of the IAEA under Research Contract No. 3694/GS.

1. INTRODUCTION

In an earlier publication [1] we reported that cotton plants treated with a ^{14}C-carbofuran formulation made up of EVA (polyethylene-polyvinyl acetate co-polymer) looked healthier than those treated with solutions (e.g. ^{14}C-carbofuran and cold carbofuran) or the control plants. The formulation had a slow release because the formulation flakes retained radioactivity (17%) even after 32 weeks. The radioactivity in the soil treated with the formulation was found to be low for 5 weeks, then increased to 52% of the applied dose at 12 weeks and finally gradually decreased. More radioactivity was recovered from the plant leaves than from the stem and roots. The highest radioactivity in the soil was observed from the 10 cm diameter samples at a depth of up to 13 cm.

The aim of the present investigation was to study the fate and magnitude of the released carbofuran insecticide from one half dose of ^{14}C-carbofuran formulation (0.5 g/pot) in order to establish the efficacy of a ^{14}C-carbofuran formulation with EVA for pest control in cotton crops grown under local conditions.

2. MATERIALS AND METHODS

The sample ^{14}C-carbofuran formulation made up of EVA with a specific activity of 0.0169 mCi/g (supplied by the Gesellschaft für Strahlen- und Umweltforschung mbH, Munich) that was used in our previous experiment [1] was applied similarly in the present studies with the modification that the dose per pot was reduced by half.[1] Twenty-five plastic pots with a diameter of 30 cm and lined with polyethylene film were filled with 22 kg of good homogenized soil. To 12 of these prepared pots, 0.5 g of formulation flakes (1.5 to 2 cm) containing 9.6 mg of carbofuran and with 8.45 μCi radioactivity were applied in a circle with a 5 cm diameter and at a depth of 5 cm. Four cotton seeds of the variety NIAB-78 were placed in the centre of each pot. Cotton seeds were planted in the remaining pots to produce the control plants. A single plant was maintained in each pot and the other plants removed from each pot were analysed separately to measure their radioactivity. First irrigation was applied 2 days after germination. Normally, the pots were irrigated twice a week for the first 5 weeks and then weekly. Sampling was done every 30 days (in duplicate) for cotton plants, soil and also for the recovered formulation flakes. Soil samples were collected with the help of mini-soil samplers from the 10 cm diameter samples (which were again divided into an upper 13 cm and a lower 13 cm portion), and from the 10 to 15 cm and the 15 to 30 cm diameter samples. This was done to obtain information about the migration rate of carbofuran

---

[1] 1 Ci = 3.70 × 10^{10} Bq.

TABLE I. PEST INCIDENCE (%) ON 2 MONTH OLD COTTON PLANTS (CONTROL AND TREATED)

Type of pests	Control	Formulation treatment
White fly	8	0
Jassids	2	0
Thrips	12	4
Aphids	56.4	16.8
Spotted bollworm	13	2.3
Pink bollworm	3.3	0

from the polymeric matrix to the soil and its further movement in the soil when applied in low amounts. The radioactivity of these samples was measured by combustion in a Packard sample oxidizer and by liquid scintillation counting [2]. The pest incidence data were recorded following the standard methods of screening for aphids (*Aphis gossypci* Glov.), jassids (*Emrasca devastans* Dist.), thrips (*Thrips tabaci* Lind.), spotted bollworm (*Earias insula* Biosd. and *E. fabia* Stoll.), pink bollworm (*Pectinophora gossypiclla* Saund.) and white fly (*Bemisia tabaci* Gen.) [3].

3. RESULTS AND DISCUSSION

The plants were studied for the incidence of pests at the ages of 2 and 4 months (Tables I and II). The results indicated that 2 month old cotton plants (Table I) treated with the carbofuran formulation showed better resistance against different pests than the 4 month old treated plants (Table II) when compared with the control plants. Although aphid attack was reduced when the plants were 4 months old, the treated plants showed a higher per cent incidence of aphids than at 2 months.

It is evident from Table III that the radioactivity in the formulation flakes fell to almost one-half when the plants were 1 month old and then decreased gradually with the increase in time. It is believed that the release of pesticide from the formulation depends on irrigation water; the young plants were irrigated more frequently than the older ones. The radioactivity in the soil increased gradually with the increase in time. Two to three month old plants showed more radioactivity than the 4 month old plants, probably because the older leaves had started to fall and few leaves remained on the plant. The total per cent recovery at each sampling was at least

TABLE II. PEST INCIDENCE (%) ON 4 MONTH OLD COTTON PLANTS (CONTROL AND TREATED)

Type of pests	Control	Formulation treatment
White fly	0	0
Jassids	0	0
Thrips	0	0
Aphids	34.8	30.5
Spotted bollworm	0	0
Pink bollworm	2	1.5

TABLE III. PER CENT RECOVERY OF ^{14}C-CARBOFURAN AT VARIOUS TIMES FROM THE WHOLE COTTON PLANT, TOTAL POT SOIL AND RECOVERED FORMULATION FLAKES

Time (months)	Soil	Plant tissue	Recovered formulation flakes	Total recovery (%)
1	24.7	15.5 (6.7 + 8.8)[a]	55.5	95.7
2	27.5	24.3 (16.5 + 7.8)[a]	45.1	96.9
3	38.6	22.1 (13.6 + 8.5)[a]	36.6	97.4
4	41.60	23.0 (10.4 + 12.6)[a]	31.7	96.3

[a] Per cent recovery of ^{14}C-carbofuran from extra cotton plants (2 or 3) removed from pots after 3 weeks of germination for the purpose of maintaining a single plant in each pot.

TABLE IV. PER CENT RECOVERY OF ^{14}C-CARBOFURAN FROM DIFFERENT SOIL ZONES AND FROM DIFFERENT PLANT PARTS AT VARIOUS TIMES WHEN APPLIED IN THE SOIL IN THE FORM OF CONTROLLED RELEASE FORMULATIONS

Time (months)	0–10 cm (upper 13 cm)	0–10 cm (lower 13 cm)	10–15 cm	15–30 cm	Root	Stem	Leaves
1	14.6	3.9	4.9	1.3		Whole plant 15.5	
2	16.6	4.3	1.4	5.2		Whole plant 24.3	
3	22.3	6.0	4.0	6.3	3.4	2.7	8.4
4	30.5	3.3	0.70	7.1	2.9	2.3	5.2

96%; the rest may have been volatilized or some experimental error may have occurred. Table IV shows the release of ^{14}C-carbofuran from the formulation into the soil, its further movement in the soil and its uptake by different plant parts. Like our earlier findings [1], the highest radioactivity in the soil was observed from the 10 cm diameter samples up to a depth of 13 cm and it increased with the increase in time. The reason is that this area was in the immediate vicinity of the carbofuran formulation. The outermost zone of the soil (15 to 30 cm diameter) also showed a gradual increase in radioactivity with time, probably because the small amounts of pesticide which were not lost by volatilization or taken up by the plant rose with the irrigation water towards the edge of the pot. Root, stem and leaf samples could only be analysed separately when the plants were 3 months old. More radioactivity was recovered from the leaves than the roots and stem, which is in agreement with our previous findings [1]. Four month old plant tissues contained less radioactivity, which may be due to the fact that plants grown in pots are under stress and show early senescence, leaves start falling and the plant weight is reduced. This may be one of the reasons why 4 month old plants offer less resistance against pests (Table II).

The present technique of applying low amounts of carbofuran formulation (0.5 g/pot) instead of 1 g/pot applied previously definitely resulted in some plant protection, but was not as effective as the higher dose [1]. Initially, the cotton plants looked healthier (up to 3 months) but then their resistance decreased in comparison to that of the control plants. It is suggested that the half dose applied in the present studies is lower than desirable and that probably two-thirds of the dose could better serve the purpose.

ACKNOWLEDGEMENTS

The authors wish to thank the International Atomic Energy Agency for supporting the work and the Pakistan Atomic Energy Commission, Islamabad, for providing the research facilities. Thanks are also due to M. Bahadir and his group at the Gesellschaft für Strahlen- und Umweltforschung mbH, Munich, for providing the ^{14}C-carbofuran formulation with EVA.

REFERENCES

[1] JAMIL, F.F., QURESHI, M.J., BASHIR, N., BAQVI, S.H.M., "Studies on the controlled-release carbofuran formulation for pest control in cotton using isotope techniques", Proc. 13th Int. Symp. on Controlled Release of Bioactive Materials, Norfolk, VA (1986) 238–239.

[2] SMITH, G.N., LUDWIG, P.D., WRIGHT, K.C., BAURIEDEL, W.R., Laboratory Training Manual on the Use of Nuclear Techniques in Pesticide Research (1983).
[3] BHATTI, M.A., "Evaluation and selection of insect resistant varieties/strains of cotton", Ten Years of NIAB, Report on Research and Other Activities, Nuclear Institute for Agriculture and Biology, Faisalabad (1982) 39–44 (Internal report).

IAEA-SM-297/10

FATE OF PESTICIDES IN SOILS OF VARIOUS ORGANIC MATTER

L. HORVÁTH, F. KLING, L.P. SIMON
Institute of Isotopes of the Hungarian Academy of Sciences,
Budapest, Hungary

Abstract

FATE OF PESTICIDES IN SOILS OF VARIOUS ORGANIC MATTER.
  The degradation of ^{14}C-parathion, ^{14}C-carbofuran and ^{14}C-EPTC was studied in laboratory experiments using garden mould, sandy soil and sandy soil amended with humic acid. The degradation rates for all three pesticides were highest in the garden mould. The intermediate products of oxidative degradation were separated by thin layer chromatography.

1. INTRODUCTION

  The variability in the physical and biological properties of soil gives rise to imponderables in the prediction of the fate of pesticides. Degradation is a complicated function of volatilization, leaching, biotic and abiotic conversions.
  Various aspects of the environmental behaviour and the fate of pesticides have been investigated. The influence of the soil–water content on the organophosphorus pesticide degradation has been studied by Ou et al. [1] and Saltzman et al. [2]. Several authors have reported the results of studies made on the role of microorganisms in pesticide degradation [3–5]. The non-living component of the organic content, e.g. humic acid, is capable of forming complexes with pesticides [6]. In the experiments reported on here we have made an attempt to differentiate between the effects of the living and non-living components of the soil organic content.

2. MATERIALS AND METHODS

  Carbon-14-parathion (0.0-diethyl-0-p-nitro U-^{14}C phenyl phosphorothionate) (specific activity: 28.15 mCi/mmol), ^{14}C-carbofuran (2.2-di-^{14}C methyl-2.3-dihydro-3-^{14}C benzofuranyl-7-N-methyl carbamate) (specific activity: 21.38 mCi/mmol), and ^{14}C-EPTC (S-ethyl-N, N-di-n-propyl ^{14}C-carbonyl thiocarbamate) (specific activity: 5.5 mCi/mmol) were prepared in our laboratory.[1] Non-radioactive parathion and carbofuran were purchased from Supelco,

---

[1] 1 Ci = 3.70 × 10^{10} Bq.

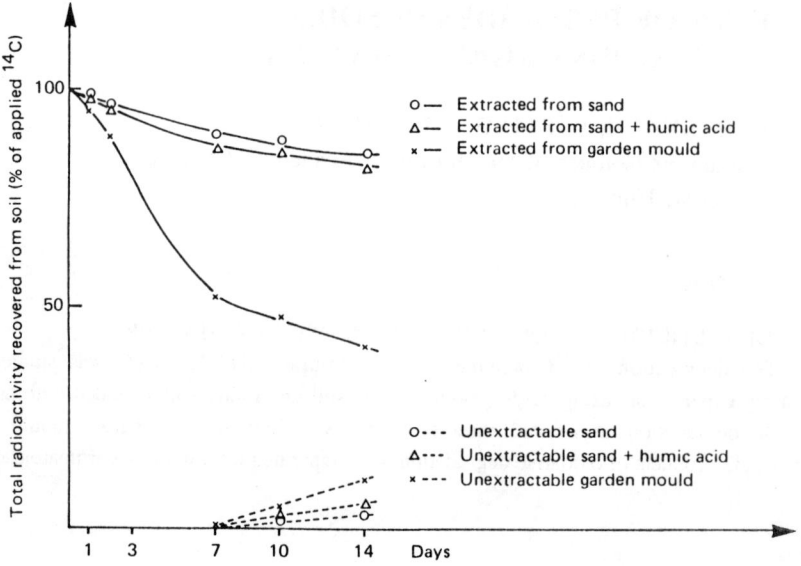

FIG. 1. *Depletion of ^{14}C-parathion from treated soils.*

Sacramento, United States of America. Soil samples (250 g) were placed in 500 mL Erlenmeyer flasks. The soils included freshly collected garden mould (pH 5.9, organic matter 32.9%), sandy soil (pH 5.3, organic matter 1.22%) and sandy soil amended with humic acid (pH 6.5, organic matter 5.2%). Humic acid was isolated from brown coal.

The soil samples were treated with the labelled pesticides at the 5 ppm level. Two series of samples were treated with unlabelled parathion and carbofuran, respectively. The Erlenmeyer flasks were kept at a constant temperature of 25°C without a stopper.

Three 10 g soil samples were withdrawn and extracted three times each with an acetone–methanol–benzene (1:1:1) mixture. The extracts were combined and filtered through Whatman No. 42 filter paper, the solvent was evaporated and the volume was adjusted to 10 mL. Aliquots of the extracts were quantitated by a liquid scintillation counter and analysed by thin layer chromatography. The distribution of the radioactivity on the chromatogram was detected using a thin layer scanner and the radioactive spots were scraped off and quantitated by the liquid scintillation method.

After extraction with the organic solvent mixture, the ^{14}C activity remaining in the extracted soil was determined by combustion in the O_2 flow and measurement of the evolved $^{14}CO_2$ in a gas proportional counter.

Non-radioactive parathion and carbofuran were applied at the same rate in parallel experiments. The extracted soils were analysed using the radiometric

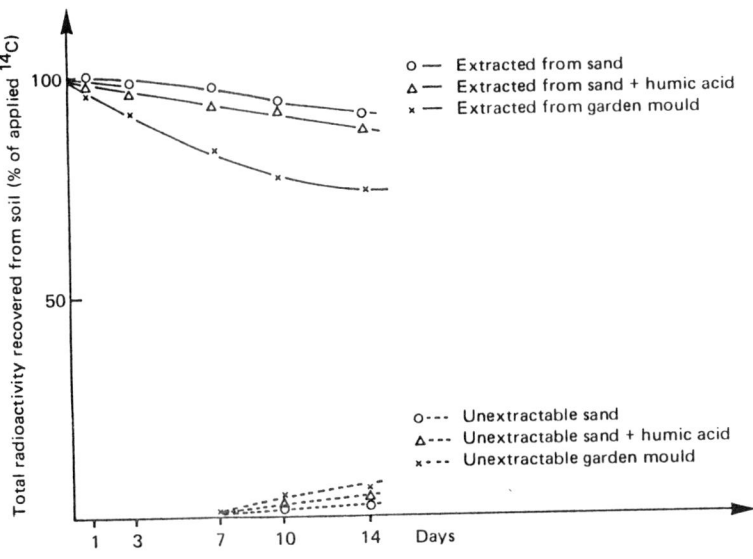

FIG. 2. *Depletion of ^{14}C-carbofuran from treated soils.*

enzymatic method in which tritium labelled acetylcholine is used [7] to determine whether the unextractable residue contained any anti-cholinesterase compounds.

3. RESULTS AND DISCUSSION

The dissipation of all three pesticides was quicker in garden mould than in the low organic content soil samples (Figs 1–3). This is inevitable and can be attributed to the metabolic activities of the soil microorganisms, although other factors also contribute to the loss of parent compounds.

EPTC showed the highest rate of depletion. This can be attributed to the higher rate of metabolism and to the high vapour pressure. The recovery of EPTC is higher in the humic acid treated soil. This supports the findings of Worobey and Webster [6], who established that humic substances can bind pesticides.

The unextractable residues were higher in garden mould for each of the three pesticides. No measurable amount of activity remained bound to the sand in the EPTC experiment (Fig. 3).

We investigated whether the bound residues possessed any anti-cholinesterase activity. Using the radiometric enzymatic method the acetylcholinesterase enzyme blocking capability of the extracted soils was measured. Neither parathion nor carbofuran experiments showed active residues higher than 0.01 ppm. This indicates that no significant fraction of bound residue is in a chemical form with anti-cholinesterase activity.

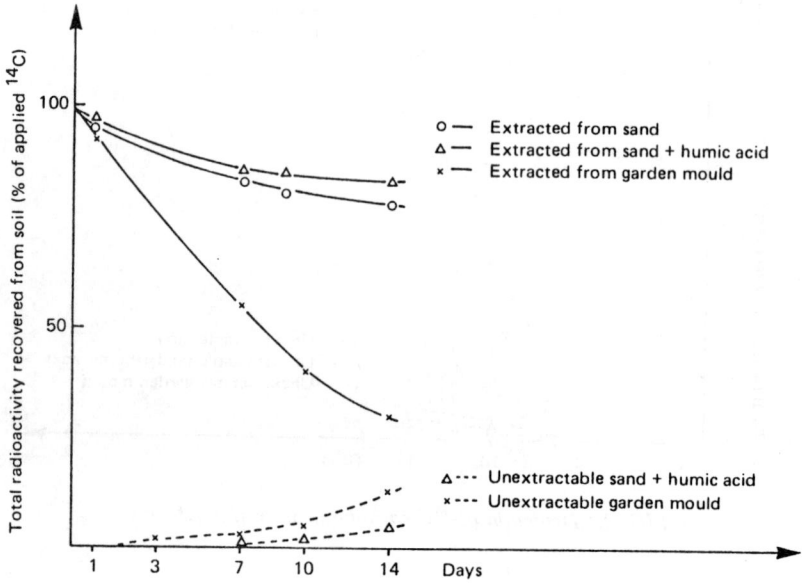

FIG. 3. Depletion of ^{14}C-EPTC from treated soils.

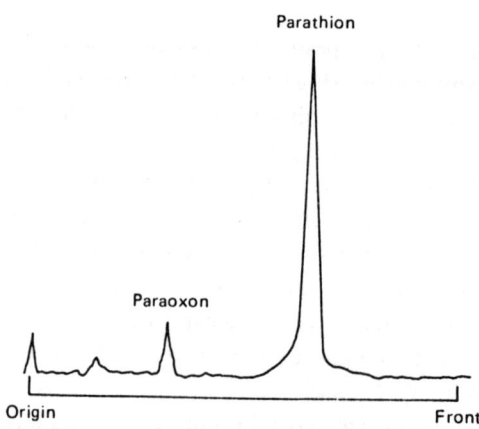

FIG. 4. Thin layer chromatogram of the extract from ^{14}C-parathion treated garden mould.

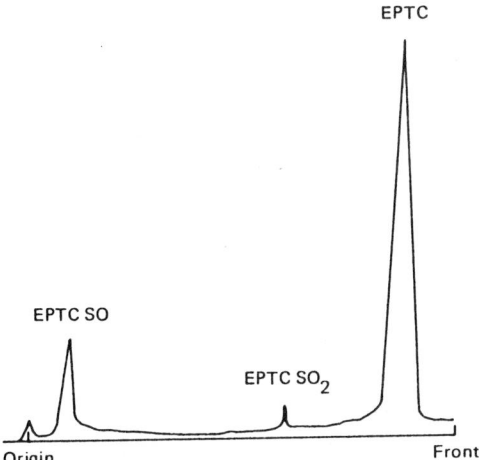

FIG. 5. *Thin layer chromatogram of the extract from ^{14}C-EPTC treated sandy soil + humic acid.*

Thin layer chromatographic analyses (Figs 4 and 5) of the extracts showed that the higher dissipation rate was due to higher oxidative metabolic conversions (Tables I and II).

Enzymes which catalyse the conversion of pesticides fall into two classes: hydrolases and oxigenases [8]. Both parathion and EPTC tend to degrade by oxidation followed by hydrolysis [9, 10]. If the conditions are favourable for oxidation, the oxidative products will be detectable.

Depletion of a pesticide from a treated soil comprises leaching, evaporation and degradation. The first two are purely physical phenomena and are governed by the physico-chemical properties of the compound and the soil characteristics (i.e. volatility, solubility, absorption capacity, water content, etc.).

The series of chemical conversions which finally lead to the breakdown of the pesticide is called degradation. Such conversions may proceed on abiotic pathways due to the light, temperature and pH, or the inorganic chemical composition of the soil. Some of the conversions are carried out by the microorganisms of the soil, either by metabolizing the compounds or by the catalytic effects of the free enzymes originating from the living organism. These factors have a complicated interrelation and make it difficult to predict the fate of pesticides in soil.

The experimental results reported here support the idea that the fate of pesticides in soils is dependent on the chemical and biological composition of the soil. Nevertheless, the pesticide 'directions for use' usually distinguish soils according to their mineral composition and do not take into account the level of the organic contents.

TABLE I. THIN LAYER CHROMATOGRAPHIC ANALYSIS OF EXTRACTS OF ^{14}C-PARATHION TREATED SOILS

Rf	Radioactivity of spots				Applied radioactivity
	31 (paraoxon)		65 (parathion)		
	(counts/s)	(%)	(counts/s)	(%)	(counts/s)
Sandy soil	1520	2.2	62 506	93.9	66 551
Sandy soil + humic acid	1363	2.1	60 210	93.4	64 460
Garden mould	5095	7.8	59 306	91.1	65 044

Merck silica gel 60.
Solvent: n-hexane–acetone (4:1).

TABLE II. THIN LAYER CHROMATOGRAPHIC ANALYSIS OF EXTRACTS OF ^{14}C-EPTC TREATED SOILS

Rf	Radioactivity of spots						Applied radioactivity
	89 EPTC		10 EPTC-SO		61 EPTC-SO$_2$		
	(counts/s)	(%)	(counts/s)	(%)	(counts/s)	(%)	(counts/s)
Sandy soil	40 320	73.2	3211	5.8	—	—	55 030
Sandy soil + humic acid	42 152	80.8	5533	10.6	1055	2.0	52 118
Garden mould	42 511	80.0	6335	11.9	—	—	53 123

Merck silica gel 60.
Developing solvent: n-heptane–acetone (6:1).

It is suggested that with further experimental work simple testing methods could be developed to quantify the biotic and abiotic degradation capabilities of soils. With such methods the dose rates of certain soil applied pesticides could be better adapted to the actual soil and climatic conditions.

REFERENCES

[1] OU, L.T., RAO, P.S.C., DAVIDSON, J.M., Methyl parathion degradation in soil: Influence of soil-water tension, Soil. Biol. Biochem. **15** 2 (1983) 211.

[2] SALTZMANN, S., MINGELGRIN, U., YARON, B., Role of water in the hydrolysis of parathion and methyl parathion on kaolimite, J. Agric. Food Chem. **24** (1976) 739.

[3] FERRIS, I.G., LICHTENSTEIN, E.P., Interactions between agricultural chemicals and soil microflora and their effects on the degradation of ^{14}C-parathion in a cranberry soil, J. Agric. Food Chem. **28** (1980) 1011.

[4] HARRIS, C.R., CHAPMAN, C., HARRIS, C., TU, C.M., Biodegradation of pesticides in soil: Rapid induction of carbamate degrading factors after carbofuran treatment, J. Environ. Sci. Health **B19** (1984) 1.

[5] KARNS, J.S., MULBRY, W.W., NELSON, J.O., KEARNEY, P.C., Metabolism of carbofuran by a pure bacterial culture, Pestic. Biochem. Physiol. **25** (1986) 211.

[6] WOROBEY, B.L., WEBSTER, G.R.B., Hydrolytic release of tightly complexed 4-chloroaniline from soil humic acids, J. Agric. Food Chem. **30** (1982) 161.

[7] HORVÁTH, L., "Organophosphorus and carbamate residues in water", Methods of Enzymatic Analysis, Vol. 12, 3rd edn (BERGMEYER, H.U., Ed.), VCH, Weinheim (1986) 406.

[8] KARNS, J.S., MULDOON, M.T., MULBRY, W.W., DERBYSHIRE, M.K., KEARNEY, P.C., "Use of microorganism and microbial systems in the degradation of pesticides", Biotechnology in Agricultural Chemistry (LE BARON, H.M., MUMMA, R.O., HONEYCUTT, R.C., DUESING, J.H., Eds), ACS Symposium Series No. 334, American Chemical Society, Washington, DC (1987) 157.

[9] GOMAA, H.M., FAUST, S.D., "Chemical hydrolysis and oxidation of parathion and paraoxon in aquatic environments", Fate of Organic Pesticides in the Aquatic Environment, Advances in Chemistry, Series 111 (GOULD, R.F., Ed.), American Chemical Society, Washington, DC (1972) 189.

[10] HORVÁTH, L., PULAY, Á., Metabolism of EPTC in germinating corn: Sulfone as the true carbamoilating agent, Pestic. Biochem. Physiol. **14** (1980) 265.

IAEA-SM-297/34

FIELD EVALUATION OF CONTROLLED RELEASE CARBOFURAN FORMULATIONS IN INDONESIAN FLOODED RICE*

R.M. WILKINS
Department of Agricultural and Environmental Science,
University of Newcastle upon Tyne,
Newcastle upon Tyne,
United Kingdom

Haeruddin TASLIM, Hendarsih SUHARTO
Sukamandi Research Institute for Food Crops,
Subang, West Java,
Indonesia

Abstract

FIELD EVALUATION OF CONTROLLED RELEASE CARBOFURAN FORMULATIONS IN INDONESIAN FLOODED RICE.

Few of the uses of pesticides, and other biologically active substances, in the open environment achieve their potential efficiency. From an understanding of the dynamics of the active agent in the environment, some of the benefits of controlled delivery methods in reducing losses can be obtained. For soil applied pesticides, the use of polymeric matrix granular systems could be economic. Granules based on kraft lignins have been shown to protect and extend the release of active agents under field conditions in tropical flooded rice. As part of this process, two types of lignin based granules were evaluated in the field in flooded rice in West Java, Indonesia, the objective being to compare their performances with conventional granules. A soil applied systemic insecticide, carbofuran, was used. Natural infestations of insects and viruses were monitored in this randomized block small plot trial throughout the season. Yields, and other crop growth metameters, were evaluated. The most important pest pressures came from leaf- and planthoppers, with viruses absent. Under these conditions grain yields with the experimental lignin granules were 22% above the control (3.18 t/ha; variety Pelita I/1), compared with an increase of 7.5% with the conventional formulation.

1. INTRODUCTION

Lignins can be used in controlled delivery systems for pest management in both chemical bonding and physical incorporation techniques [1]. The formation of a physical matrix of the active agent with a water insoluble alkali lignin has been well

---

* Research carried out in association with the IAEA under Research Agreement No. 3826/CF.

reported [2] and is the basis of herbicidal products [3, 4]. Granular formulations of other pesticides have been prepared [5], particularly for the soil applied systemic insecticide and nematicide, carbofuran. This compound is much used in tropical agriculture, especially in vegetables and in rice, where it is active against a wide range of important insect pests [6].

The compatability of the pesticide with lignin, and thus its ability to form a coherent matrix, largely depends on the similarity of the respective solubility parameters. Formulations based on carbofuran can be prepared under melt conditions with pine kraft lignin [1]; these can be moulded or extruded, or cooled and granulated. Such methods yield granules with a minimum carbofuran content of 25%. Lower active ingredient (a.i.) content can be achieved by the inclusion of polar modifiers. Additionally, the incorporation of other polymers can improve, for example, abrasion resistance and release profiles through water soluble coreleasers [7].

Two granular formulations of carbofuran based on kraft lignin have been evaluated in tropical flooded rice. The most effective ways of using such slow release materials in the crop were investigated under Philippine conditions [8]. To evaluate the potential benefits, including improved effiency, the lignin granules were field tested under a range of locations, with varying climatic, edaphic and pest conditions. We report here the results of field testing performed in flooded rice in West Java, Indonesia.

2. MATERIALS AND METHODS

2.1. Formulations

The matrix was prepared under melt conditions [7], using technical grade carbofuran (FMC Corporation, Philadelphia, USA) and pine kraft lignin (Indulin AT, Westvaco, North Charleston, SC, USA). This was cooled, granulated and sieved to pass a 1.0 mm aperture but not 0.5 mm. Two formulations were prepared: CLFG20 containing 20% carbofuran, 30% polyethyleneglycol and 10% vinyl polymer; and CLF45 containing 45% carbofuran and 10% vinyl polymer. The standard formulation was Furadan 3G, containing 3% carbofuran on a sand base.

2.2. Treatments

The granules were broadcast on to the soil surface following field preparation and prior to rice transplanting. The treatments were individually pre-weighed, mixed with sand to give the same amount per plot as for the standard and applied by hand. The granules were then incorporated using a rototiller to a depth of 5 cm. One application rate (0.6 kg a.i./ha) was used.

2.3. Field experiments

The field experiment was established in the wet season 1984–1985 on the Sukamandi Experimental Farm using small plots (100 m^2) separated by low mud walls with water inlet and outlet flows isolated [9]. A complete randomized block design with six replicates was used. Rice seedlings (variety Pelita I/1; 20 days old) were transplanted (25 cm by 25 cm spacing) on 25 December 1984; normal crop management included fertilizer, weed control and water level inputs but no insect control.

2.4. Assessments

2.4.1. Insects

Observations were made weekly on 20 hills in each plot starting at three weeks after transplanting (WAT) until harvest. Insects counted included the rice whorl maggot (*Hydrellia philippina*), green leafhopper (*Nephotettix* spp.), white backed planthopper (*Sogatella furcifera*), brown planthopper (*Nilaparvata lugens*), rice gall midge (*Orseolia oryzae*), rice black bug (*Scotinophora coarctata*), stink bug (*Leptocorisa oratorius*) and stem borers (by count of dead hearts and white heads).

2.4.2. Crop condition and growth

For each plot the height of the plants and the number of productive tillers per hill were determined prior to harvest. Any symptoms of diseases were noted. At harvest, 1 m around each plot was discarded, the crop harvested by hand, machine threshed and moisture contents and 1000 grain weights measured. The trial was harvested on 12 April 1985.

3. RESULTS AND DISCUSSION

3.1. Insect pest populations

3.1.1. Green leafhopper

The green leafhopper (*Nephotettix* spp.) was observed in the field trial from 5 weeks after transplanting (WAT). The mean counts of insects knocked on to sticky boards placed around the plants for 20 hills in each plot are given in Table I. All of the treatments reduced the number of green leafhopper (GLH) below that of the control up to 10 WAT, with the lignin formulations having lower populations than the standard. To compare the overall performance of the three formulations, the total

TABLE I. THE MEAN NUMBERS OF GLH ON 20 HILLS PER PLOT AT DIFFERENT WEEKS AFTER TRANSPLANTING (WAT): SUKAMANDI, WET SEASON 1985

Treatment	WAT								Total	% reduction
	4	5	6	7	8	9	10	11		
Control	2.8a	17.5a	20.2a	18.5a	27.2a	27.5a	20.0a	4.3a	186a	—
Fur.3G	2.5a	11.8b	11.3b	10.8b	13.2b	17.2ab	16.0b	4.3a	87b	36
CLFG20	3.7a	11.0b	12.5b	16.0ab	17.3ab	15.0b	14.8b	1.7b	92b	33
CLF45	0.8a	8.0b	10.0b	13.3ab	9.8b	11.2b	15.8b	2.2ab	71c	49

Application rate: 0.6 kg a.i./ha.
In a column, means followed by a common letter are not different at P = 0.05. Duncan's Multiple Range Test, data transformed $\sqrt{x + 0.5}$.

GLH pressure is represented in Table I by the total insect count per plot. In this case the CLF45 granule was the only treatment to significantly reduce season long GLH numbers below the other plots. In addition to weekly comparisons, the overall reduction of insect pests is an indication of the increased efficiency of the slow release method.

3.1.2. White backed planthopper

The white backed planthopper (*Sogatella furcifera*) occurred in greater numbers than GLH in the field trial. The weekly mean numbers in 20 hills in each plot are shown in Table II. Although both lignin formulations consistently had lower white backed planthopper (WBPH) populations than either the control or the standard these differences were not significant until 8 WAT, indicating an extension of carbofuran action late in the season. These differences had disappeared for both lignin granules by 10 WAT but for the standard granule by 8 WAT.

Generally, although WBPH numbers were higher than GLH (see Section 3.1.1), control by carbofuran was less effective than for GLH and this trend was reflected in the responses to the controlled release formulations. For overall WBPH control, the two lignin granules generated similar reductions in the pest numbers, both significantly different from the control and the standard Furadan 3G.

3.1.3. Other insect pests

The results for the rice whorl maggot (*Hydrellia philippina*) at 3 WAT, the rice black bug (*Scotinophora coarctata*) at 13 WAT and the stink bug (*Leptocorisa oratorius*) at 13 WAT are shown in Table III. Differences between control and treated are evident early in the season (3 WAT) but not at the end of the growing period. Similar low numbers were observed for the rice gall midge (*Orseolia oryzae*) at 11 and 13 WAT. The other insects observed were too few to obtain reliable data.

3.2. Crop growth and yields

3.2.1. Crop growth

The growth of the crop, measured prior to harvesting, is shown in Table IV. No virus diseases were observed; the plots treated with carbofuran were more uniform and ripened 5 days earlier than the controls. Although no significant difference ($P = 0.05$) exists, the plots treated with the controlled release formulations were taller, and had more productive tillers per hill, than the standard treatment. This was consistent with earlier results from the Philippines [8].

TABLE II. THE MEAN NUMBERS OF WBPH ON 20 HILLS PER PLOT AT DIFFERENT WEEKS AFTER TRANSPLANTING (WAT): SUKAMANDI, WET SEASON 1985

Treatment	WAT									Total	% reduction
	4	5	6	7	8	9	10	11			
Control	23.2a	30.3a	39.5a	31.2a	24.8a	23.2a	11.7a	2.3a		186a	—
Fur.3G	12.5b	18.0b	24.8b	19.3b	25.5a	20.8a	11.5a	3.0a		136ab	27
CLFG20	8.2b	20.0b	24.7b	16.5b	16.3b	14.5b	11.7a	2.0a		114b	39
CLF45	9.3b	21.3b	21.8b	18.5b	17.1ab	16.8ab	11.5a	2.0a		118b	37

In a column, means followed by a common letter are not different at $P = 0.05$. Duncan's Multiple Range Test, data transformed $\sqrt{x + 0.5}$.

TABLE III. INFESTATION BY OTHER INSECT PESTS IN CARBOFURAN GRANULE TRIALS IN FLOODED RICE: SUKAMANDI, WET SEASON 1985

Treatment	Whorl maggot[a]	Black bug[b]	Stink bug[c]
Control	16.7a	2.3a	1.8a
Furadan 3G	7.2b	2.8a	1.8a
CLFG 20	6.5b	2.8a	1.5a
CLF45	6.5b	2.8a	1.2a

Application rate: 0.6 kg a.i./ha.
In a column, means followed by a common letter are not different at $P = 0.05$. Duncan's Multiple Range Test, data transformed $\sqrt{x + 0.5}$.
a: number of infested hills per plot at 3 WAT.
b: number of black bugs per 20 hills per plot at 13 WAT.
c: number of stink bugs per 20 hills per plot at 13 WAT.

TABLE IV. CONTROLLED RELEASE CARBOFURAN GRANULES FIELD TRIAL. RICE CROP GROWTH MEASURED PRIOR TO HARVEST: SUKAMANDI, WET SEASON 1985

Treatment	No. of productive tillers per hill	Plant height (cm)
Control	6.13a	103.87b
Furadan 3G	6.42a	110.13a
CLFG 20	6.85a	112.70a
CLF45	6.92a	113.03a

Application rate: 0.6 kg a.i./ha.
In a column, means followed by a common letter are not different at $P = 0.05$. Duncan's Multiple Range Test.

TABLE V. CONTROLLED RELEASE CARBOFURAN FIELD TRIAL.
GRAIN YIELDS AT HARVEST: SUKAMANDI, WET SEASON 1985

Treatment	Plot yields (kg/64 m^2)			t/ha	% compared with control
	Crude	Dried	Cleaned		
Control	25.72a	21.38a	20.35a	3.18	—
Furadan 3G	26.90a	23.03a	26.20a	3.42	7.5
CLFG 20	30.93a	26.20a	24.80a	3.88	22.0
CLF 45	30.79a	26.01a	24.83a	3.89	22.3

Application rate: 0.6 kg a.i./ha.
In a column, means followed by a common letter are not different at $P = 0.05$. Duncan's Multiple Range Test. Grain dried to 14% moisture content.

TABLE VI. CONTROLLED RELEASE
CARBOFURAN FIELD TRIAL. RICE GRAIN
QUALITY: SUKAMANDI, WET SEASON 1985

Treatment	Moisture content at harvest (%)	1000 grain weight (g)
Control	23.59a	28.18a
Furadan 3G	23.50a	28.59a
CLFG 20	23.26a	29.28a
CLF 45	23.23a	28.77a

Application rate: 0.6 kg a.i./ha.
In a column, means followed by a common letter are not different at $P = 0.05$. Duncan's Multiple Range Test (moisture content values transformed to arcsin $\sqrt{x}$).

3.2.2. Grain yields

The central 64 m^2 (1024 hills) of each plot was harvested by hand, weighed, cleaned and dried; the results are presented in Table V. The two lignin formulations gave similar increased yields: an increase of 22% above the control compared to a 7.5% increase with the standard Furadan 3G. These increases were probably associated with suppression of GLH and WBPH populations, although these pests were at low densities. In the absence of rice virus diseases the potential yield increases were limited (compare yield responses of over 100% in the presence of rice tungro virus [8]), but do demonstrate the value and increased effectiveness of controlled release insecticide formulations.

3.2.3. Grain quality

Each plot harvested was sampled and the moisture contents and 1000 grain weights were measured. The results (Table VI) indicated a slight improvement in early maturity (lower moisture content) and in grain filling (higher 1000 grain weight) compared with the standard formulation at the same dosage rate.

ACKNOWLEDGEMENTS

The authors acknowledge the financial support for this study from the Tropical Development and Research Institute, London.

REFERENCES

[1] WILKINS, R.M., Br. Polym. J. **15** (1983) 177.
[2] THOMPSON, H.E., ALLAN, G.G., NEOGI, A.N., Int. Pest Control **23** (1981) 10.
[3] EPA label for Forest-Aid, Greenshield, Seattle, USA, EPA Reg. No. 43740-1.
[4] WILKINS, R.M., BLACKMORE, T. "Extended availability of propachlor", Proc. British Crop Protection Conference — Weeds-1987, British Crop Protection Council, Thornton Heath, Surrey (1987) 679.
[5] WILKINS, R.M., Pestic. Sci. **15** (1984) 258.
[6] PATHAK, M.D., Int. Pest Control **8** (1968) 4.
[7] CHANSE, A., WILKINS, R.M., "Use of lignins in polymeric controlled release systems", Proc. Cellucon 1986 — Wood and Cellulosics, Wrexham, UK, July 1986, Ellis Horwood, Chichester, UK (in press).
[8] WILKINS, R.M., BATTERBY, S., HEINRICHS, E.A., AQUINO, G.B., VALENCIA, S.L., J. Econ. Entomol. **77** (1984) 495.
[9] HEINRICHS, E.A., et al., Manual for Testing Insecticides on Rice, International Rice Research Institute, Manila, Philippines (1981).

DEVELOPMENT AND TESTING OF PESTICIDE FORMULATIONS WITH THERMOPLASTIC POLYMERS*

F. KORTE, G. PFISTER, M. BAHADIR
Institut für Ökologische Chemie,
Gesellschaft für Strahlen- und
　Umweltforschung mbH,
Munich, Federal Republic of Germany

Abstract.

DEVELOPMENT AND TESTING OF PESTICIDE FORMULATIONS WITH THERMOPLASTIC POLYMERS.

　Thermoplastic polymers, mainly ethylene-vinyl acetate co-polymers (EVA), were used to prepare controlled release formulations of pesticides. EVA films with the incorporated herbicides desmetryn and atrazine successfully suppressed weed growth in field trials with white cabbage and sweet corn, respectively, at amounts of application comparable or even lower than those necessary with conventional techniques. The transfer of desmetryn from polymer films into the vapour phase was determined in the laboratory and found to be negligible compared with the release into water coming into contact with the films. Other examples of experimental applications of EVA formulations were control of aquatic weeds with the herbicides terbutryn and simetryn, root zone application of carbofuran in rice and cotton cultivation and the combined attraction and elimination of tsetse flies.

1. INTRODUCTION

　Agricultural production as well as storage of harvested crops worldwide are dependent on the use of various types of pesticides.

　The Food and Agriculture Organization of the United Nations assumes that the rapidly increasing demand for food, particularly in developing countries, will support the future growth of pesticide use despite intensive parallel efforts to introduce alternative methods of pest control such as biological and integrated systems [1]. The hazards of this development, as currently foreseen, are the increasing pesticide residues in agricultural products, the contamination of surface and even groundwater and the breakdown of natural ecosystems, which may lead to the destruction of additional food sources. Increasing risks of toxicity to humans who come into direct contact with pesticides during their application can also be expected.

---

* Research carried out in association with the IAEA under Research Agreement No. 3471/CF.

Inefficient delivery to the target site caused by conventional application techniques, degradation processes and physical transport phenomena usually makes it necessary to apply excess pesticide in order to guarantee the required effects. This can result not only in an increased environmental burden, but also in growing production costs. Attempts to solve these problems have been made in different ways, e.g. new insecticides such as synthetic pyrethroids have been developed which have a reduced mammalian toxicity, need lower amounts of application and are degraded more easily in the environment after use. Waste of the active ingredients, however, caused by inadequate application methods must be avoided by other means. New methods of pesticide formulation could help towards this aim. In the ideal case, such formulations should allow better control of pesticide release in duration and amount and should also be able to protect the active ingredients against premature degradation processes and physical displacement until they are released. In practice, this will be possible only in combination with a close adaptation of such formulations to the requirements of special problems in pest management because of the divergent needs in this field.

In our own studies, which were part of the international research programme on controlled release pesticides, co-ordinated by the Joint FAO/IAEA Division of Isotope and Radiation Applications of Atomic Energy for Food and Agricultural Development in Vienna, we formulated pesticides with thermoplastic polymers such as polyethylene, LDPE, and ethylene-vinyl acetate co-polymers, EVA. The effectiveness of some of these formulations in various applications was investigated in field trials and biological tests. Some of these studies are reported here.

2. RESULTS AND DISCUSSION

Polymer films are widely used in agriculture and horticulture to protect crops from adverse weather influences, thus enabling an earlier harvest and higher and more constant yields. In field trials lasting over three vegetation periods, we studied the effectiveness of weed control by means of EVA cover and mulch films with incorporated herbicides compared with conventional ground spraying of the same chemicals. The herbicidal effects were determined by the evaluation of locally growing weeds and sown indicator plants. The herbicide desmetryn was applied in this way in test plots with white cabbage as the crop [2]. Because of the lack of experience with these film formulations the design of the field trials had to be somewhat modified from one vegetation period to the next. Comparisons between the different annual results are therefore inherently difficult.

Figure 1 shows typical examples of the efficiency of weed control in the 1986 trial. The desmetryn released from the EVA cover films markedly suppressed weed growth, but for comparable results the same amounts of application as those used with conventional treatments were needed. The almost quantitative elimination of

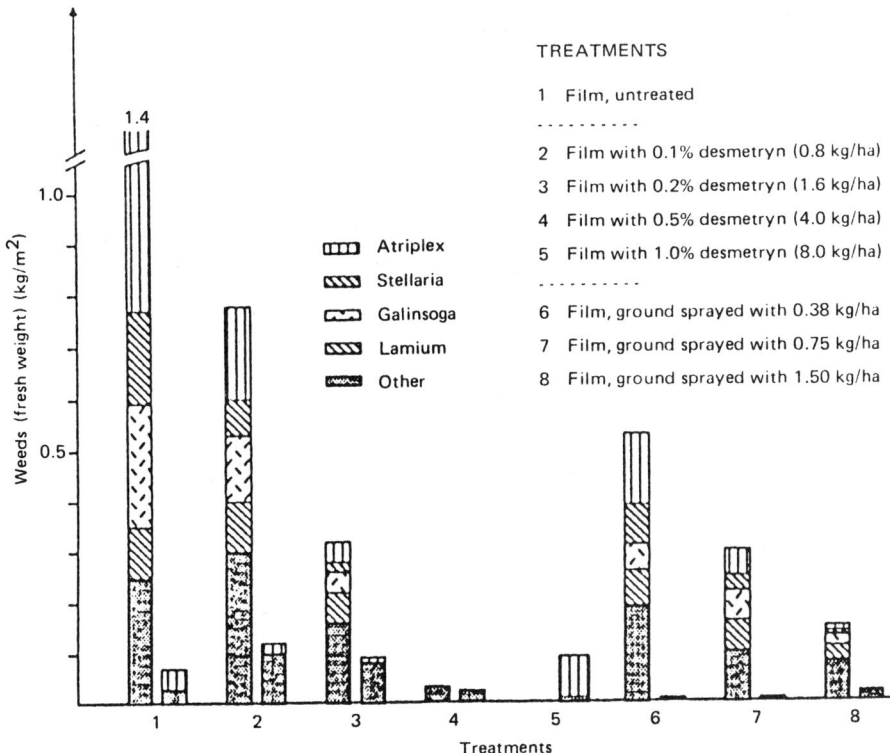

FIG. 1. Development of weeds in white cabbage test plots under cover (left bars) and mulch sheets (right bars) (1986).

weeds throughout all treatments, when mulch sheets were applied, must be mainly due to the higher temperatures and mechanical suppression under these films. The herbicidal effect on the indicator plants (Table I) with different sensitivities seems to be higher with some species in the film treatments. On the whole the results are comparable to those obtained with the natural weeds.

The crops were free of residues, with isolated exceptions. Residues in the soil were generally lower in treatments with the controlled release formulations than when the same quantity of desmetryn was conventionally applied.

The treatment of sweet corn with mulch films in which atrazine had been incorporated was very effective against natural weed growth as well as against indicator plants even at the lowest herbicide concentration of 0.025% in the film, corresponding to 0.2 kg/ha compared with 0.72 kg/ha, which is the amount of application usually recommended. The control plots treated by ground spraying, however, showed a comparable suppression of weed growth, which may be due to the generally higher activity of atrazine under the films.

TABLE I. DEVELOPMENT OF INDICATOR PLANTS AFTER 8.5 WEEKS (mean fresh weight in gram) (1986)

Treatment	Cress	Salad	White cabbage	Kohlrabi	Mixed weeds
1: film untreated	439	5	435	245	3372
2: film with 0.1% desmetryn (0.8 kg/ha)	0	5	925	448	2430
3: film with 0.2% desmetryn (1.6 kg/ha)	0	12	375	245	1306
4: film with 0.5% desmetryn (4.0 kg/ha)	0	0	65	6	154
5: film with 1.0% desmetryn (8.0 kg/ha)	0	0	0	0	53
6: film, ground sprayed with 0.38 kg/ha	510	0	790	519	2316
7: film, ground sprayed with 0.75 kg/ha	268	27	890	655	2263
8: film, ground sprayed with 1.50 kg/ha	29	1	1173	551	1484

The release of the active ingredients from the films in the field trials already described usually lasted for a period of 3 to 4 weeks. In the ideal case the whole amount of the released pesticide should be available for the soil–plant system to be treated. The application of the film formulations, however, raises the question of possible undesired losses from the upper side of the films to the atmosphere. To obtain information about the transfer of active ingredients into the vapour phase, it was necessary to develop a method which also allowed the determination of the

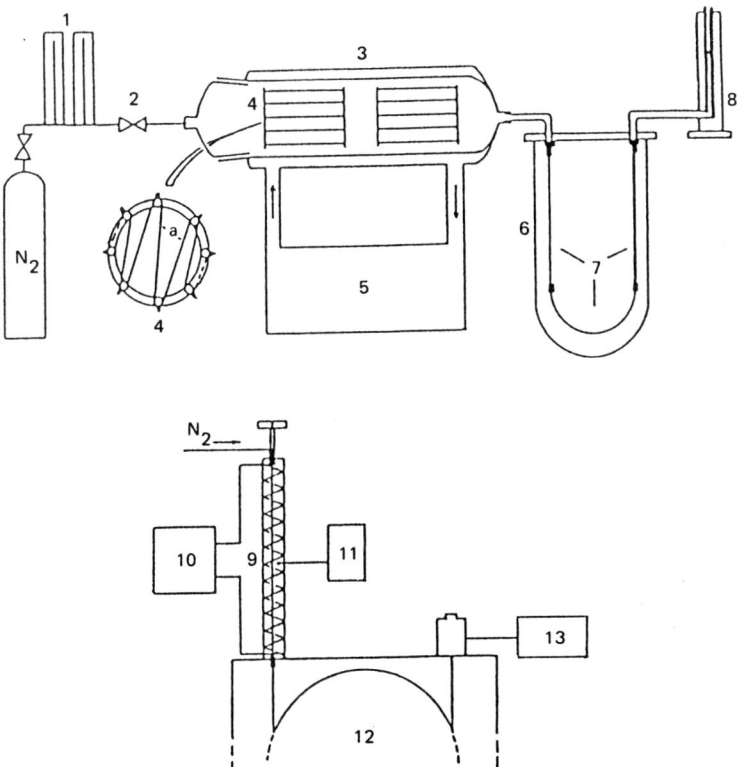

FIG. 2. Apparatus for measuring migration of bioactive substances from polymer films into the vapour phase: (1) gas drying tubes; (2) needle valve; (3) thermostated glass cylinder; (4) cylindrical glass cages covered with film formulations (a); (5) thermostat; (6) Dewar flask (T − 25°C); (7) fused silica capillary column (530 μm, 3 × 20 cm) coated with methyl silicone; (8) flow meter; (9) capillary column in heating tube; (10) voltage regulator; (11) thermocouple; (12) capillary GC/FID; (13) integrator.

evaporizing fraction of low volatile pesticides from films under controlled conditions (Fig. 2) [3]. Table II gives some data obtained with 70 μm thick films with incorporated desmetryn, which has a vapour pressure of 1.3×10^{-4} Pa [4]. It is obvious that the rate of evaporation is very low and does not increase much even at elevated temperatures of up to 75°C compared with a thin crystalline layer of the compound on glass plates with the same surface ratio. To obtain a better approximation to natural atmospheric conditions films were also hung inside a fume hood with an air exchange rate of 600 m^3/h, and periodically analysed for their herbicide concentration. Under these conditions 72 to 82% of the desmetryn were released in 61 days. Studies in pure water (Fig.3) gave a complete release within 6 hours. These

TABLE II. TRANSFER OF DESMETRYN FROM EVA FILMS AND COATED GLASS PLATES INTO DRY NITROGEN (flow 0.7 L/h)

Concentration of desmetryn in films (%)	Temperature (°C)	Normalized release (pg·L^{-1}·cm^{-2})	Amount of desmetryn released after 4 days (% of initial value)
0.07	25	2.9	0.0083
0.14	0	8.5	0.014
	25	9.1	0.015
	50	11.2	0.018
	75	14.5	0.024
0.35	25	12.9	0.011
0.35 washed	25	12.5	0.010
0.62	0	19.4	0.0070
	25	21.0	0.0075
	50	24.8	0.0090
	75	26.4	0.0095
Glass plates			
= 0.07	25	13.3	0.038
= 0.62	0	204.0	0.069
	25	222.2	0.075
	50	330.0	0.111
	75	394.0	0.132

results indicate that during field application losses to the atmosphere are negligible. Pesticide transport, apart from rainfall, mainly occurs as a result of dripping of condensed water from the lower film surface. This causes a decreasing concentration compared with the top surface, directing the main diffusion of the incorporated chemical down towards the soil–plant system (Fig. 4).

Another field of application for EVA formulations of herbicides, shaped as cords and tapes, is the control of aquatic weeds in tropical countries. Tests with such formulations of terbutryn were performed under natural conditions in Indonesia [5]. The results obtained by the local investigator cannot be reported here.

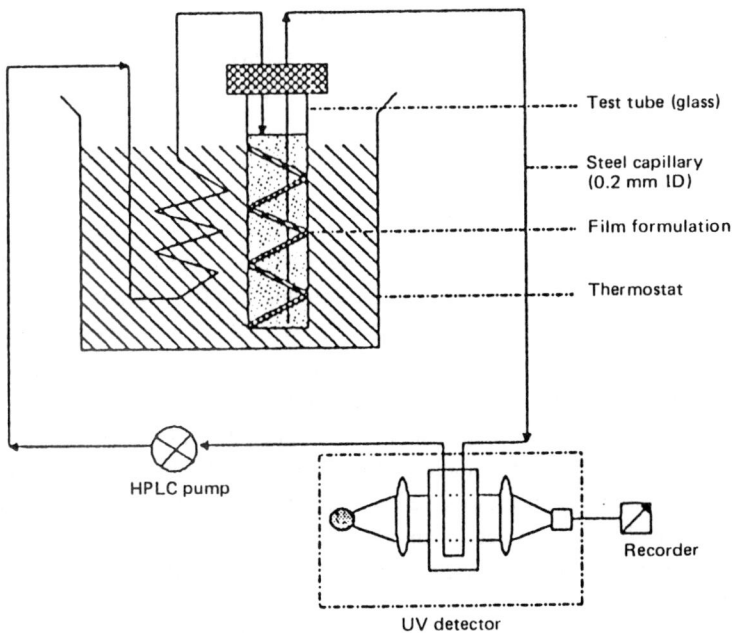

FIG. 3. Flow control measuring device for determination of pesticide release into water.

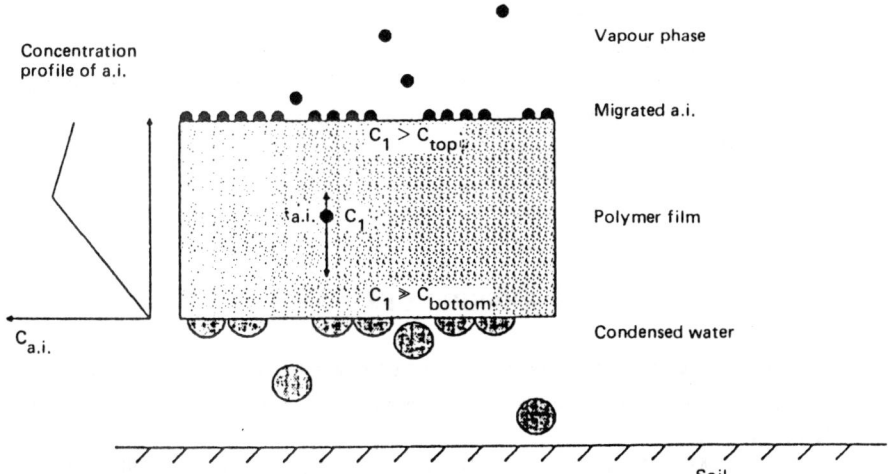

FIG. 4. Direction of diffusion of incorporated active ingredients with low vapour pressure in agricultural polymer films.

In our laboratory experiments with duckweed (*Lemna minor*) as the aquatic test plant, the ^{14}C labelled herbicide simetryn was released from an EVA cord into a flow system [6]. The uptake of the herbicide by the plant was found to be cumulative, with initial bioconcentration factors of up to 112. An effective suppression of weed growth was achieved at simetryn concentrations of 0.2 to 0.06 ppm, released over a period of 20 days.

Carbon-14 labelling was also used in studies with carbofuran which had been incorporated into EVA cords to protect this systemic insecticide from premature degradation during soil application, thus prolonging its effectiveness while minimizing environmental contamination. Field studies with this formulation applied in the root zones of rice [7] and cotton plants were performed by investigators in Hungary and Pakistan, respectively, with promising results.

At least one example for possible useful applications of EVA formulations in the control of disease vectors should be mentioned. EVA films with blue pigmentation and incorporated endosulfan have been prepared to control tsetse flies (*Glossina morsitans*). The blue colour acts as an optical attractant towards these insects, which come in contact with the insecticide when they rest on the film. In tests carried out in the IAEA Laboratory at Seibersdorf, Austria, a 100% mortality of tsetse flies was achieved within 2 to 3 days after exposure times to the film of only 10 min to 1 h. At present, investigations with this formulation, mainly to study the attractive properties, are taking place in Nigeria under the auspices of the FAO.

The examples presented here should give an idea of how versatile EVA formulations of pesticides can be used and adapted to different problems in pest management. Of course, there are some disadvantages of this method such as non-biodegradability of the polymer, limited pesticide load of the matrix and economic considerations, which may possibly lead to some limitations in its utilization. For many purposes, however, these formulations can be applied in a way that will allow recovery of the polymer matrix after depletion of the pesticide depot and its recycle into the formulation process as long as the properties of the polymer do not undergo adverse changes.

REFERENCES

[1] FOOD AND AGRICULTURE ORGANIZATION OF THE UNITED NATIONS, International Code of Conduct on the Distribution and Use of Pesticides, Annex, FAO, Rome (1986).
[2] BAHADIR, M., PFISTER, G., LORENZ, W., HERRMANN, R., KORTE, F., Z. PflKrank. PflSchutz **94** (1) (1987) 34.
[3] KRAXENBERGER, M., PFISTER, G., BAHADIR, M., J. Appl. Polym. Sci. **34** (1987).

[4] FRIEDRICH, K., STAMMBACH, K., J. Chromatogr. **16** (1964) 22.
[5] SOERJANI, M., Controlled Release Herbicides for Aquatic Weeds , Progress Report to the Joint FAO/IAEA Division, Vienna, Research Contract No. 3477/R2/GS, 1986.
[6] KAMAL, M., PFISTER, G., BAHADIR, M., LAY, J.P., J. Contr. Rel. (accepted for publication).
[7] BAHADIR, M., PFISTER, G., Chemosphere **16** (6) (1987) 1273.

SEED DRESSING WITH CONTROLLED RELEASE FORMULATIONS
Evaluation using a radioisotope technique and yield estimations for the control of aphids and stem nematodes in field beans (Vicia faba L.)

B.C. SCHIFFERS*, P. DREZE**,
J. FRASELLE*, M.C. GASIA**

* Chaire de phytopharmacie

** Chaire de physique et CAMIRA

Faculté des sciences agronomiques de l'Etat,
Ministère de l'éducation nationale
  et de la culture française,
Gembloux, Belgium

Abstract

SEED DRESSING WITH CONTROLLED RELEASE FORMULATIONS: EVALUATION USING A RADIOISOTOPE TECHNIQUE AND YIELD ESTIMATIONS FOR THE CONTROL OF APHIDS AND STEM NEMATODES IN FIELD BEANS (*Vicia faba* L.).

Previous studies have shown that the incorporation of systemic insecticides in seed coatings, designed as controlled release formulations, is a combined operation (sowing and treatment) which uses much less pesticide for the same period of activity. The carbofuran incorporated into coated field bean seeds (*Vicia faba* L.) at a rate of 3 mg active ingredient per seed resulted in a reduction of up to 96% in the stem nematode populations (*Ditylenchus dipsaci* (Fil.)(Kühn)) found in plants 4 months after sowing; 3 mg of carbofuran per seed correspond to 0.9 to 1.2 kg a.i./ha, depending on the sowing density. The long persistence of nematicide activity for such a quantity of active ingredient is obtained by the slow release of carbofuran from the seed coatings. As might be expected for a systemic pesticide incorporated into the soil, carbofuran has no effect against pollinators and pests, predators or parasites. The calibrated seeds of field beans were coated using the rolling technique. Tritiated carbofuran can be homogeneously incorporated into the matrix, or can be incorporated into a resin or encapsulated in a wide range of matrices. The controlled release effect of all these formulations has been characterized using a radioisotope technique. In a laboratory test, carbofuran was released three times more slowly from coatings than when formulated as commercial microgranules. Infestation in field beans after the growing season was determined by comparing the number of stem nematodes found in treated and untreated plants at different dates. It was found that incorporation of carbofuran into a urea–formaldehyde resin formulation provided the best protection against stem nematodes (95.8% of the control, 4 months after sowing) and aphids, and also increased the yield (+58%). Carbofuran markedly improved the yields of treated plots. The size of grains harvested on the treated plots and the amount of proteins were significantly higher than those of the untreated plots. The carbofuran residues in the flour of harvested grains, determined by gas–liquid chromatography, were always below the threshold level.

1. INTRODUCTION

Incorporation of systemic pesticides into the soil, as done with granule formulations or seed dressing with pesticides, to prevent drift and inadequate dosage problems results in a marked decrease in pollution risks in the environment [1].

Systemic insecticide granules applied over seed furrows were more effective when placed close to the seeds [2–4]. As pesticides in the soil have half-lives of only a few weeks [5], the results obtained were often better when granules were applied several times. However, this practice greatly increased the residues in harvested products, often beyond the threshold level [3, 4].

Seed dressing is prepared by combining a biologically active agent with excipients (polymer, resin, etc.) which regulate the delivery of the agent to the target. The benefits of seed dressing with controlled release formulations are potentially considerable compared with conventional formulations and application technologies. They facilitate a more precise delivery of pesticide to the appropriate target site; they preserve the biological activity at an effective level during critical periods by controlling the release of active ingredients and protecting them from premature decomposition [5]; they use much less pesticide for the same period of activity, within the limitations of the available pesticide; and they offer the advantage of combining two operations (sowing and treatment) [6, 7].

2. SEED DRESSING WITH CONTROLLED RELEASE FORMULATIONS

The calibrated seeds of field beans (*Vicia faba* L.) (cv. Exelle), with a diameter of 8.5 to 9.5 mm, were coated using the rolling technique [8–11]. Seed dressing was done in a rotating sphere. To the moistened seeds were added alternatively an aqueous solution of an adhesive and a dry mixture (clays, silicates, sawdust, etc.) with one or more adhesive materials. Further careful drying will reduce the excess water, so obtaining a hard matrix around the seeds.

To the seed coating process was added a concentrated suspension of carbofuran (Curater SC 330 flowable) diluted in a solution of adhesive. Water soluble adhesives, such as polyvinyl alcohols, combine on drying and create a network of film around the active agent. This is called a 'one step' or 'classical' (CL) formulation [12].

TABLE I. COMPOSITION OF SEED DRESSING WITH CONTROLLED RELEASE FORMULATIONS (3 mg CARBOFURAN/SEED) *(The total amount of coating material added to the field bean seeds corresponds to 30% of their own weight)*

Formulations	Clays		Absorbent	Stickers			Resin
	Bentonite (%)	Vermiculite (%)	Perlite (%)	Versicol S19 (%)	Vinarol ST (%)	Mowilith DL45 (%)	(%)
Classical (CL)	49	10	29	10	2	–	–
Urea–formaldehyde (UF) (resin titrating 4% of carbofuran)	27	14	–	4	–	9	46
Starch xanthide (SX) (resin titrating 12% of carbofuran)	50	10	17	5	–	–	18

Carbofuran can also be incorporated into a resin or encapsulated in a wide range of matrices which are ground to a suitable dimension, mixed with other pelleting adjuvants and added to the seeds. This is called a 'two step' formulation, e.g. a urea–formaldehyde resin (UF) or a starch xanthide matrix (SX).

For the laboratory tests, solutions of tritiated carbofuran (specific activity: 15.5 μCi/mg or 574 kBq/mg) in acetone with concentrations of 3.22, 5.3 and 20.7 mg/mL for CL, UF and SX formulations, respectively, were prepared for incorporation into the seeds. Tagging was achieved by adding 1 mL of the radioactive solution to 3 and 9 g of concentrated suspension (at 33% active ingredient) for CL and SX coatings, respectively, and to 3 g of crystallized carbofuran in acetone for the UF formulation, resulting in about 2 mg of carbofuran per seed.

It was found that seed coating material amounting to 30 wt% of the field bean seeds was enough to incorporate directly or indirectly 3 mg of carbofuran per seed. The details of each formulation are given in Table I.

Previous studies [13] have shown that the rolling technique is able to distribute, on average, the desired dose of carbofuran on to the seeds, with a variation from seed to seed that depends on the amount of coating materials on each seed (linear correlation coefficients: +0.85 and +0.94 for CL and UF formulations, respectively). Carbofuran degradation in the coating material does not occur until at least 12 months.

3. STANDARD WASHING TEST

To evaluate carbofuran release from the coated seeds, a standard washing test was used [13]. This involved placing five seeds on a sand bed (100 g) in a Büchner funnel with a diameter of 7.5 cm and a height of 7 cm, retaining 25 or 30 mL of distillated water in the funnel for 24 hours and then draining it off (Fig. 1). The eluted volume was noted; 1 mL of the eluate was then mixed with 10 mL of a scintillation mixture (Lumagel–Lumac) to measure the radioactivity. The results are expressed in the total weight per cent of carbofuran released from the seed coatings after washing (Table II and Fig. 2). At the end of the test, the radioactivity remaining in the seeds and the sand was measured.

To compare the controlled release of seed coatings with a commercial granular formulation of carbofuran (Curater 5G), the standard washing test was also used; the microgranules contained the same amount of carbofuran as the five coated seeds. The carbofuran released from the microgranules was also extracted from the eluted water and measured by gas chromatography. The results are expressed in the total weight per cent of carbofuran released at the end of the washing test (Table II and Fig. 2).

Some eluate samples were analysed by gas chromatography to identify the radioactivity (GC column: WCOT 20 m × 0.32 mm; phase: Sil 5 CB; temperature: programmed from 70 to 200°C; the on-column injection mode; NPSD detector; azobenzene as the internal standard).

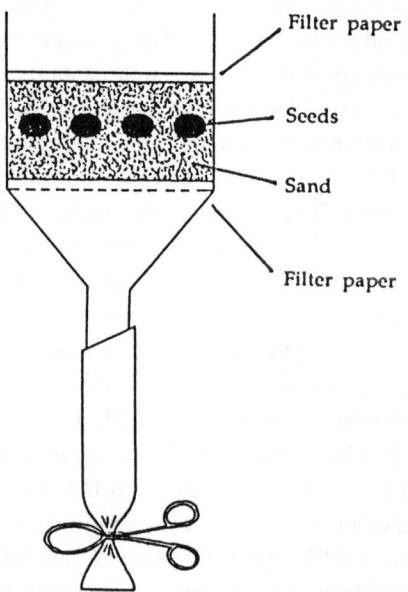

FIG. 1. Büchner funnel used for the standard washing test.

TABLE II. RESULTS OF A STANDARD WASHING TEST EXPRESSED IN PER CENT OF TOTAL WEIGHT OF CARBOFURAN RELEASED AT THE END OF THE TEST FOR EACH FORMULATION (CL: CLASSICAL COATING; SX: STARCH XANTHIDE; UF: UREA–FORMALDEHYDE)
(Average of six replicates)

Cumulated eluate volumes (mL)	CL coating (%)	SX coating (%)	Cumulated eluate volume (mL)	UF coating (%)	Microgranules (Curater 5G) (%)
24	6.4 ± 1.7	6.3 ± 1.0	30	9.0 ± 0.7	26.3 ± 2.8
50	10.3 ± 1.4	13.4 ± 1.9	60	24.1 ± 1.0	20.7 ± 1.5
75	11.1 ± 2.3	13.0 ± 1.9	90	18.9 ± 1.7	17.5 ± 1.2
100	10.7 ± 1.7	13.0 ± 1.9	120	10.4 ± 0.7	16.5 ± 1.0
125	14.9 ± 1.4	13.0 ± 1.9	150	7.5 ± 1.0	9.7 ± 0.9
150	9.9 ± 0.7	9.4 ± 1.5	180	4.8 ± 0.2	5.7 ± 0.8
175	10.4 ± 1.2	8.6 ± 1.5	210	3.4 ± 0.2	2.2 ± 0.3
200	8.4 ± 1.1	6.9 ± 1.6	240	2.5 ± 0.2	0.9 ± 0.3
225	6.5 ± 1.5	4.9 ± 1.8	270	2.0 ± 0.2	0.5 ± 0.3
250	4.8 ± 1.3	3.3 ± 1.0	300	2.0 ± 0.1	
275	3.4 ± 1.4	2.2 ± 0.7	330	1.7 ± 0.1	
300	2.3 ± 1.0	1.5 ± 0.6	360	1.3 ± 0.1	
325	0.8 ± 0.4	1.1 ± 0.4	390	1.6 ± 0.1	
350		0.7 ± 0.2	420	1.5 ± 0.3	
375		0.6 ± 0.2	450	1.4 ± 0.2	
400		0.3 ± 0.1	480	1.3 ± 0.1	
425		0.3 ± 0.1	510	1.3 ± 0.1	
450		0.3 ± 0.1	540	1.1 ± 0.1	
475		0.2 ± 0.0	570	1.0 ± 0.1	
500		0.2 ± 0.1	600	0.9 ± 0.1	
525		0.2 ± 0.1	630	2.1 ± 0.1	
550		0.1 ± 0.1			
575		0.2 ± 0.1			
600		0.2 ± 0.1			
625		0.1 ± 0.1			
650		0.1 ± 0.1			
675		0.1 ± 0.0			

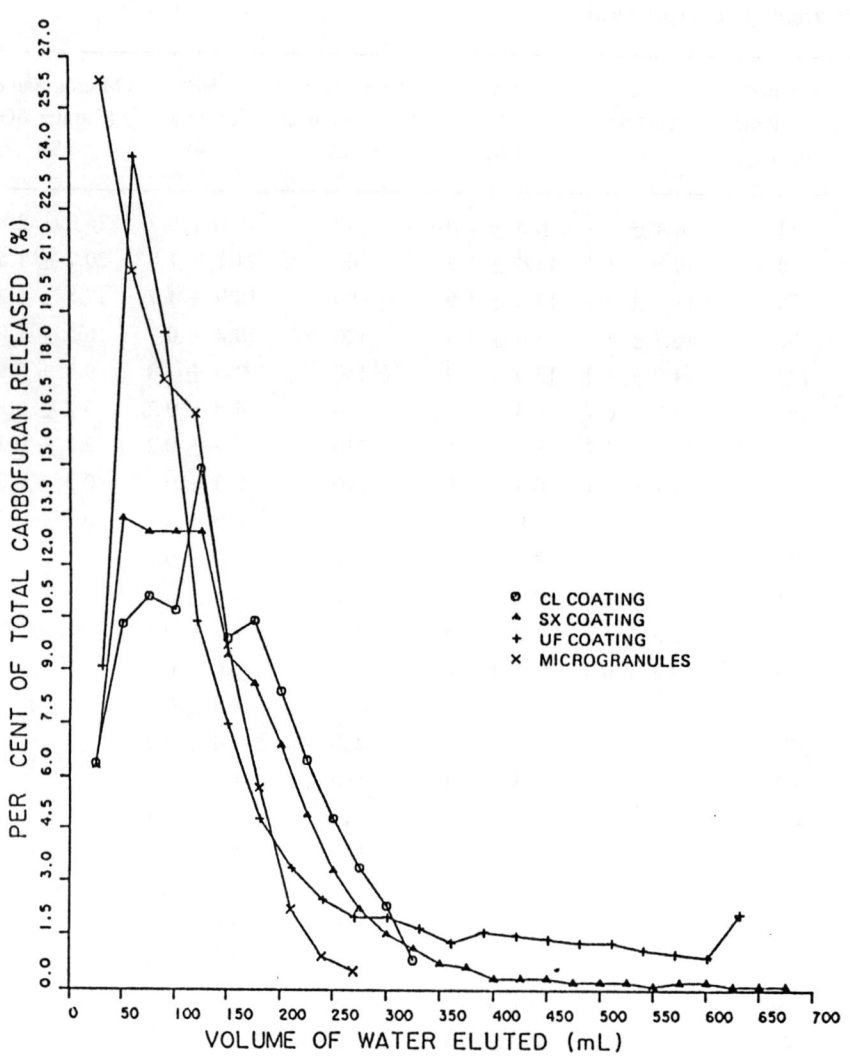

FIG. 2. Per cent of total carbofuran released from microgranules (Curator SG) and seed coatings depending on the volume of water eluted in a standard washing test.

TABLE III. PER CENT OF TOTAL AMOUNT OF CARBOFURAN REMAINING IN COATINGS WHEN COMMERCIAL MICROGRANULES ARE USED UP AND RELEASE RATIO VALUES

Formulations	Microgranules used up (%)	Volume of water needed to release % of the total amount of carbofuran			Release ratio
		50	75	95	
		(mL)			
Microgranules (Curater 5G)	0	90	120	180	1.0
CL coating	26	125	200	275	1.5
UF coating	22	120	210	540	3.0
SX coating	23	125	175	300	1.7

The release of coatings and microgranules can be compared by measuring the total amount of carbofuran remaining in the formulations when the microgranules are used up (A%), and the volume of water needed to release 50, 75 and 95% of the carbofuran. The release ratio (Rr) can be established as Rr = coatings V_{95}/microgranulesV_{95} (Table III).

The results of this evaluation indicate that seed dressing formulations appear to offer a higher retention of active agent than commercial microgranules of carbofuran, the best being the UF coating, which has an A% equal to 22% and an Rr three time greater than that of the microgranules. No significant level of radioactivity was found in the sand or the seeds.

4. CARBOFURAN DISTRIBUTION AND METABOLIZATION IN FIELD BEANS

Ten millilitres of a tritiated carbofuran solution (77 µg a.i./mL) were applied to the soil of flower pots (17.2 kBq/pot) in which a single 1 week old field bean was grown. The distribution of carbofuran in field beans was investigated by stem and leaf calcination, in sequence with a BMO–ICN calcinator. Sampling and calcination were carried out 2, 4 and 6 weeks after application. The results (Table IV) showed a very heterogeneous distribution of radioactivity in the plants: the oldest leaves, especially the edges, accumulated much more radioactivity than the youngest leaves.

TABLE IV. DISTRIBUTION OF ^{3}H RADIOACTIVITY IN FIELD BEANS 2, 4 AND 6 WEEKS AFTER APPLICATION TO THE SOIL (AVERAGE OF TWO PLANTS) *(The leaves were numbered from 1 to 10 according to their order of formation; sometimes the edges of the leaves were cut and measured separately from the rest of the leaves)*

Sample	Distribution of ^{3}H radioactivity		
	After 2 weeks	After 4 weeks	After 6 weeks
		(Bq/g dry weight)	
Stem	167	224	63
Leaf 1	175	2108	1334
Edge 1	190	–	–
Leaf 2	68	1368	1498
Edge 2	227	4332	–
Leaf 3	139	1211	1675
Edge 3	189	4066	–
Leaf 4	67	1997	–
Leaf 5	50	1536	1014
Leaf 6		1059	543
Leaf 7		775	282
Leaf 8		199	165
Leaf 9			53
Leaf 10			35

The same results were found with ^{14}C-carbofuran. These results are in agreement with those of Ashworth and Sheets in tobacco plants [14].

In another experiment, the metabolization of carbofuran in field beans was investigated. Sixty flower pots each received one field bean seed coated with ^{14}C-carbofuran (coating application with 5.3 kBq and 1.14 mg of carbofuran/pot; specificity activity: 4.6 Bq/μg of carbofuran). Sixty other pots each received one seed and were watered with 10 mL of a ^{14}C-carbofuran solution in water (drench application with 16.8 kBq and 3 mg of carbofuran/pot; specific activity: 5.6 Bq/μg of carbofuran). The control pots received only 10 mL of water. The pots were placed in a greenhouse. All the plants were watered daily with 50 mL of water. On days 17 and 35 after sowing the plants were sampled for carbofuran extraction and three plants from each type of treatment (untreated, coating application and drench application) were placed in a cage and received 20 aphids (*Acyrtosiphum pisum* Harris) per plant to determine the persistence of carbofuran activity.

Two samples taken from a 50 g plant were blended twice with 200 mL of acetone at high speed for 2 minutes and then filtered on glass filter G4. The filtrate was concentrated, extracted four times with 50 mL of chloroform in a separatory

TABLE V. PER CENT CARBOFURAN AND ITS METABOLITES IN FIELD BEANS 17 AND 35 DAYS AFTER SOWING WHEN AN ACTIVE AGENT IS APPLIED BY SEED COATING OR BY SOIL WATERING WITH A RADIOACTIVE SOLUTION *(Means are expressed in becquerels for the total sample extracted)*

Samples	Coating application			Drench application		
	Means (average of six TLC counts)	ppm (dry weight)	% of total activity	Means (average of six TLC counts)	ppm (dry weight)	% of total activity
After 17 days						
Carbofuran	1151 ± 178	65.0	38.5	177 ± 5	8.3	34.2
3-OH carbofuran	1451 ± 210	81.9	48.6	242 ± 71	11.3	46.8
3-OH glycoside	248 ± 48		8.3	58 ± 23		11.3
Unidentified	137 ± 10		4.6	40 ± 10		7.7
Total	2987		100.0	517		100.0
After 35 days						
Carbofuran	158 ± 38	5.9	12.3	2 ± 1	0.08	1.2
3-OH carbofuran	564 ± 134	21.1	43.8	47 ± 17	1.6	23.9
3-OH glycoside	488 ± 4		37.9	129 ± 36		65.3
Unidentified	77 ± 31		6.0	19 ± 3		9.6
Total	1287		100.0	197		100.0

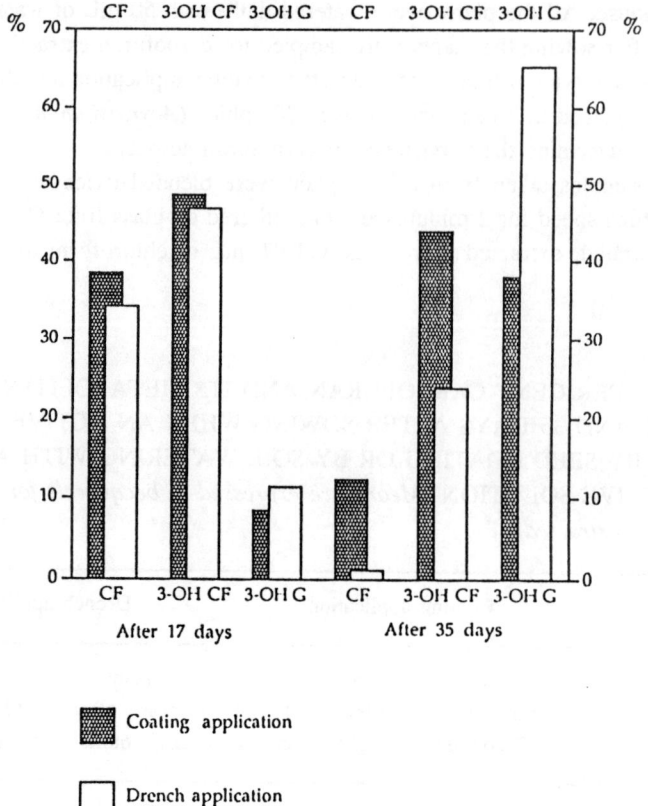

FIG. 3. *Comparative evolution of carbofuran and its metabolites in field beans 17 and 35 days after sowing when an active agent is applied by seed coating or soil watering with a radioactive solution (CF = carbofuran; 3-OH CF = 3-hydroxycarbofuran; 3-OH G = 3-hydroxycarbofuran glycoside).*

funnel and recovered in 5 mL of chloroform [15, 16]. The water containing 3-hydroxycarbofuran glycoside was refluxed for 30 minutes with 150 mL of HCl 0.25N, extracted four times with 50 mL of dichloromethane and recovered in 5 mL of dichloromethane. Carbofuran and its main metabolites (3-hydroxycarbofuran and 3-hydroxycarbofuran glycoside) were separated by thin layer chromatography (three replicates for each sample) using a mixture of chloroform–acetone–acetonitrile (4:1:1, vol./vol.). After control with a radio TLC scanner, the bands containing CF, 3-OH CF and 3-OH glycoside were scraped and directly measured by liquid scintillation chromatography. The results are given in Table V and illustrated in Fig. 3.

Seventeen days after sowing, no difference was found between the amounts of carbofuran and metabolites in the plants, regardless of the type of treatment. Also, aphids failed to establish on the plants, except for the control. Thirty-five days after sowing, higher amounts and concentrations of carbofuran and 3-hydroxycarbofuran were determined in coating application plants than in drench application plants (Table V). This explains the failure of aphids to establish on seed coating plants; the evolution of aphid populations was similar for drench application plants and the control.

5. FIELD EXPERIMENT

The experiment was of a randomized block design, consisting of four treatments (untreated and three formulations: CL, UF, SX) and four blocks. Each plot had fifty 1.5 m rows that were 0.25 m apart. Field bean seeds were sown on 3 May 1985 in each row at a sowing density of 40 seeds/m^2. Each plot was then subdivided: 25 rows were set aside for estimation of yield and in the other 25 rows field beans were sampled for nematode extractions. After 117 days withering of the field beans prevented multiplication of the stem nematodes.

For nematode extraction, the stem bases of 10 field beans collected at random were cleaned, chopped up, weighed and extracted by the mixer centrifugal flotation technique described in Ref. [17] and adapted to our laboratory conditions [18, 19].

Infestation of the field beans after different periods of growth was measured; the effects of CL, UF and SX formulations (with 3 mg carbofuran/seed) on the number of stem nematodes per gram of plant tissue at 52, 75, 97 and 117 days after

TABLE VI. EFFECTS OF VARIOUS SEED COATING FORMULATIONS (3 mg CARBOFURAN/SEED) ON THE NUMBER OF STEM NEMATODES RECOVERED FROM YIELD BEANS AT DIFFERENT TIMES AFTER SOWING

Formulations	Average number of stem nematodes/g of stem tissue (days after sowing)				Control (%)
	52	75	97	117	
Untreated	3.6 ± 1.5	11.1 ± 1.8	59.6 ± 38.4	266.9 ± 59.3	0
CL coating	0.4 ± 0.3	1.6 ± 0.7	4.6 ± 2.0	19.9 ± 8.7	92.5
UF coating	0.1 ± 0.1	0.4 ± 0.3	2.4 ± 1.6	11.2 ± 6.0	95.8
SX coating	0.5 ± 0.5	1.1 ± 0.6	22.5 ± 3.4	42.1 ± 16.4	84.3

TABLE VII. INFLUENCE OF DIFFERENT FORMULATIONS OF CARBOFURAN ON THE YIELD, THOUSAND GRAIN WEIGHT (TGW) AND PROTEIN CONTENT OF A FIELD BEAN CROP *(Average of four replicates)*

Formulations	Yield (g/m^2)	TGW (g)	Protein content (%)
Untreated	402.8[a]	320.7[a]	24.2[a]
CL coating	604.7[b]	397.5[b]	25.7[b]
UF coating	636.4[b]	412.9[b]	26.1[b]
SX coating	563.9[b]	387.0[b]	25.7[b]

[a, b] Distribution of means following the Newman and Keuls method for alpha = 0.05%.

sowing are shown in Table VI. The efficacy of each nematicide formulation was determined by comparing the percentage of infestation control (C%) at the last extraction date. This is expressed by C% = 100 − (X/Y) × 100, where X is the number of stem nematodes found in the treated plants at the last extraction date and Y is the number of stem nematodes found in the untreated plants at this date (Table VI).

The highest number of stem nematodes was found in the first observation of untreated plants and it increased rapidly. A good control was achieved during the first 3 months with all formulations of carbofuran. At the last observation (117 days after sowing) the SX formulation permitted some nematode penetration.

Half plots were harvested on 10 September 1985 (130 days after sowing). Field bean grains were cleaned and air dried before being weighed. Carbofuran markedly improved the yield (P <0.001), the thousand grain weight and the protein content; all were significantly higher than those in the untreated plots (Table VII). All the values given in the table are significantly different (P <0.05).

After extraction, the carbofuran residues in the flour of the harvested grains were determined by gas–liquid chromatography [19]. The residues attained 0.11 ppm or 0.38 ppm for plants treated with the CL or the UF formulation, respectively.

6. CONCLUSIONS

The controlled release performance of coating and microgranule formulations has been characterized using a radioisotope technique. In a laboratory test, carbofuran was released three times more slowly from coatings than when formulated as commercial microgranules.

Accumulation of carbofuran and its metabolites occurs in the oldest parts of the plants. Carbofuran release from seed coatings provides a higher amount of chemical in the plant tissues, which increases the efficacy time of the active agent.

Schiffers et al. [18, 19] demonstrated that carbofuran incorporated in seed coating formulations is an effective nematicide. In the field experiment described here, the chemical markedly reduced the number of stem nematodes found in plants 4 months after sowing (Table VI). Three milligrams of carbofuran per seed correspond to 0.9 to 1.2 kg a.i./ha, depending on the sowing density. The long persistence of nematicide activity for such a quantity of active ingredients is obtained by the slow release of carbofuran from seed coatings, as characterized in a standard washing test.

Seed dressing with the UF formulation provides the best protection against stem nematodes (95.8% of the control, 4 months after sowing) and the best yield (+58%). The SX formulation has a shorter activity than the others, but infestation is nevertheless prevented during the sensitive blossoming period (until 15 July). Therefore, yields of the SX formulation are not significantly smaller than those of the other treatments. The residues of carbofuran at harvest were always below the threshold level (0.50 ppm), but the difference in residue levels between the CL and the UF formulations illustrates the risk of prolonging the release period for too long.

ACKNOWLEDGEMENTS

The authors would like to thank P. Derenne, Plant Breeding Station, Gembloux, for helpful discussions, N. Dethier, E. Donis, B. Rezette and L. Jaumin for technical assistance, C. Vincinaux (Bayer Belgium SA), G. Pilate (vermiculite and perlite), and Allied Colloids, Belgium. Thanks are also due to the Institut pour la recherche scientifique en industrie et en agriculture, Belgium, for financial support.

REFERENCES

[1] HEDIN, P., J. Agric. Food Chem. **30** 2 (1982) 201–215.
[2] WHITEHEAD, A.G., TITE, D.J., FRASER, J.E., FRENCH, E.M., Ann. Appl. Biol. **93** (1979) 213–220.
[3] WHITEHEAD, A.G., TITE, D.J., FRASER, J.E., Ann. Appl. Biol. **103** (1983) 291–299.
[4] WINDRICH, W.A., Crop Prot. **4** 4 (1985) 458–463.
[5] SUETT, D.L., Crop Prot. **5** (1986) 165–169.
[6] McFARLANE, N.R., PEDLEY, J.B., Pestic. Sci. **9** (1978) 411–424.

[7] SCHIFFERS, B.C., Ann. Gembloux **91** (1985) 31-37.
[8] JEFFS, K.A., CIPAC Monograph 2, Heffers Printers Ltd, Cambridge, UK (1978) 99 pp.
[9] LONGDEN, P.C., ADAS Quart. Rev. **18** (1975) 73-80.
[10] FRASELLE, J., SCHIFFERS, B., Med. Fac., Landbouww. Rijksuniv., Gent **47** 2 (1982) 665-673.
[11] SCHIFFERS, B., CORNET, D., FRASELLE, J., BALANDI, Mboka-Unda, Parasitica **38** 2 (1982) 55-63.
[12] SCHIFFERS, B.C., FRASELLE, J., Ann. Gembloux **88** (1982) 165-175.
[13] SCHIFFERS, B.C., FRASELLE, J., GASIA, M., DREZE, P., "Seed dressing with controlled release formulations: Preparation and evaluation using radioisotope techniques", Research and Development of Controlled Release Technology for Agrochemicals Using Isotopes (Proc. Sem. Vienna, 1985), IAEA-TECDOC-404, IAEA, Vienna (1087) 40-41.
[14] ASHWORTH, R.J., SHEETS, T.J., J. Econ. Entomol. **63** 4 (1970) 1301-1304.
[15] BAYOUMI, O.C., SCHIFFERS, B., FRASELLE, J., Med. Fac., Landbouww. Rijksuniv., Gent **48** 4 (1983) 1007-1014.
[16] BAYOUMI, O.C., DREZE, P., Bull. Rech. Agron., Gembloux **19** (1984) 155-163.
[17] COOLEN, W.A., HENDRICKX, J., GOORIS, J., D'HERDE, C.J., Min. Agr. St., Agric. Res. Cent., Gent (1984) 9 pp.
[18] SCHIFFERS, B.C., FRASELLE, J., HUBRECHT, F., JAUMIN, L., Med. Fac., Landbouww. Rijksuniv., Gent **49** 2b (1984) 635-641.
[19] SCHIFFERS, B.C., FRASELLE, J., JAUMIN, L., Med. Fac., Landbouww. Rijksuniv., Gent **50** 3a (1985) 797-807.

IAEA-SM-297/30

Invited Paper

CURRENT TRENDS IN PESTICIDE USAGE IN SOME ASIAN COUNTRIES
Environmental implications and research needs

M. SOERJANI
Centre for Studies of the Environment
 and Human Resources,
University of Indonesia,
Jakarta, Indonesia

Abstract

CURRENT TRENDS IN PESTICIDE USAGE IN SOME ASIAN COUNTRIES: ENVIRONMENTAL IMPLICATIONS AND RESEARCH NEEDS.

The trends in pesticide consumption in some Asian countries clearly indicate that pesticides are not only a critical input in pest management, mainly to increase crop production, but also for public health, recreation, forestry, etc. However, there is also growing public concern over the impact of pesticides on public health in particular and on the quality of the environment in general. A study was made to collect information from nine countries in Asia; they are listed in order of their pesticide consumption: India, the Republic of Korea, Indonesia, Malaysia, Pakistan, Thailand, the Philippines, Sri Lanka and Bangladesh. In 1985 the total pesticide consumption of these countries was approximately 1.1 million tonnes, 70% of which (or 0.75 million tonnes) were manufactured locally. Twenty-one technical materials were produced in eight countries (excluding Sri Lanka). The total pesticide market of all these countries was estimated at US $1189 million, of which 74.2% are shared by four countries: India (26%), the Republic of Korea (19.6%), Indonesia (16.8%) and Malaysia (11.8%). The most important crop in Asia is rice, and the main pests are stem borers and planthoppers. Other important pests occur in the following crops: cotton, rubber, oilpalm, tea, sugarcane, jute, bananas, pineapples and vegetables. The problems of pesticide use are: (1) the pesticide treadmill, the increasing pesticide application and the decreasing efficacy, which increase the complexity of future pest problems, and the unequal sharing of benefits and environmental risks among the various community groups; (2) the side effects on non-target organisms and the environment, and on public health and poisoning problems during production, storage, distribution, mixing and application, as well as contamination of resources and products; (3) the lack of appropriate information for those concerned with pesticide distribution and usage; and (4) legislation problems, violation of pesticide codes and enforcement problems. Research needs, the results of which are expected to improve future public policies and actions, concern the efficacy and efficiency of pesticide application, the fate of pesticides in the environment, the development of controlled release formulations, the mode of resistance of pests and crops, the selection and breeding of crops and animals, and the development of a better integrated pest management approach towards pest problems.

1. INTRODUCTION

Most countries in Asia have economies that are based on agriculture and the current trends point towards even further development of the agricultural sector with a view to self-sufficiency, import substitution and increased exports [1]. The consequent development of technology has led to use of modern inputs of energy such as high yielding varieties of crops, fertilizers, pesticides and other agrochemicals, as well as agrophysical devices. Increasing interest and recognition are being shown by governments and private sectors in the use of pesticides to increase crop production. Therefore, pesticides are still considered a critical input to improve crop yields and to prevent crop losses before and after harvest [2].

Over the past three decades much research and numerous studies have been undertaken on pest problems in order to optimize crop yields through appropriate pest management. Efforts have been made to develop integrated pest management (IPM) and other measures of crop protection to achieve optimum sustainable yields, but use of pesticides still remains the most convenient and widely accepted method of pest control.

Along with the increasing use of pesticides, there is also a growing trend towards awareness by the public of the impact of pesticides on public health and the environment.

This paper attempts to review some of the available information concerning pesticide production and usage in some Asian countries, the consequent impact on public health and the environment, the efforts being made to develop appropriate pesticide management including pesticide regulation policies, and identification of the research needs.

2. PESTICIDE PRODUCTION AND USAGE

2.1. Important pest problems in Asia

Since pesticides are produced and used for pest control in crops and other man made environments, there is a need to briefly review the occurrence of important pests in Asia in order to illustrate the need for pesticide production and usage in Asia.

As briefly covered in the report of Gaston [1], rice is the most important crop in Asia, since it is the staple diet in most countries. The main pests associated with rice are stem borers and planthoppers. Of lesser importance are some bacterial and fungal diseases, e.g. sheath blight and stemrot [3], while banyard grass and purple nutsedge are important weeds [4]. Cotton is one of the main crops in India and Pakistan; the bollworm complex and aphids are the major pests. Rubber and oilpalm are the basic plantation crops in Malaysia and Indonesia and weeds, e.g. alang-alang and Siam, are the most prevalent problem. Tea is an important plantation crop in

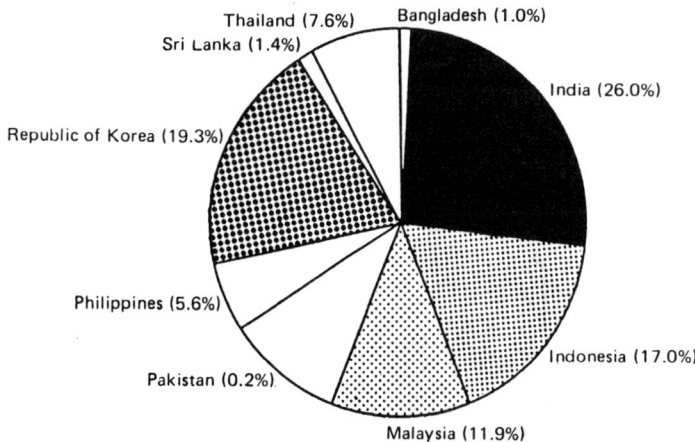

FIG. 1. *Pesticide market by country in 1985 (in millions of US $). Of these figures, insecticides account for 62.5% of the regional market, followed by herbicides with 21.4% and fungicides with 15.1% [1].*

Sri Lanka, India, China and Indonesia, and again weeds pose the greatest problem. Jute is a major export crop in Bangladesh. Sugarcane is an important plantation crop in the Philippines and Indonesia and some problems exist with disease and weeds. Bananas and pineapples are the most important plantation crops in the Philippines; they are affected by sigatoka disease and weeds, respectively. Vegetables are also among the most important crops grown in Thailand and the Republic of Korea and the diamond back moth, pod borers, aphids and cut worms are some of the common pests [1]. Water hyacinth and molesting salvinia are the two important weed problems in water resources in most Asian countries, while *Mimosa pigra* is an important aquatic weed in Thailand [5]. Mosquitoes are the most important vectors of malaria and dengue fever and use of pesticides is necessary for their control.

2.2. Pesticide consumption

A survey on pesticide usage was conducted in 1985 by the Regional Network on Pesticides for Asia and the Pacific in nine countries in Asia: Bangladesh, India, Indonesia, Malaysia, Pakistan, the Philippines, the Republic of Korea, Sri Lanka and Thailand. It revealed that the total pesticide market was estimated at US $1189 million, of which India's share was 26%; the Republic of Korea, 19.6%; Indonesia, 16.8%; and Malaysia, 11.8% (see Fig. 1) [1].

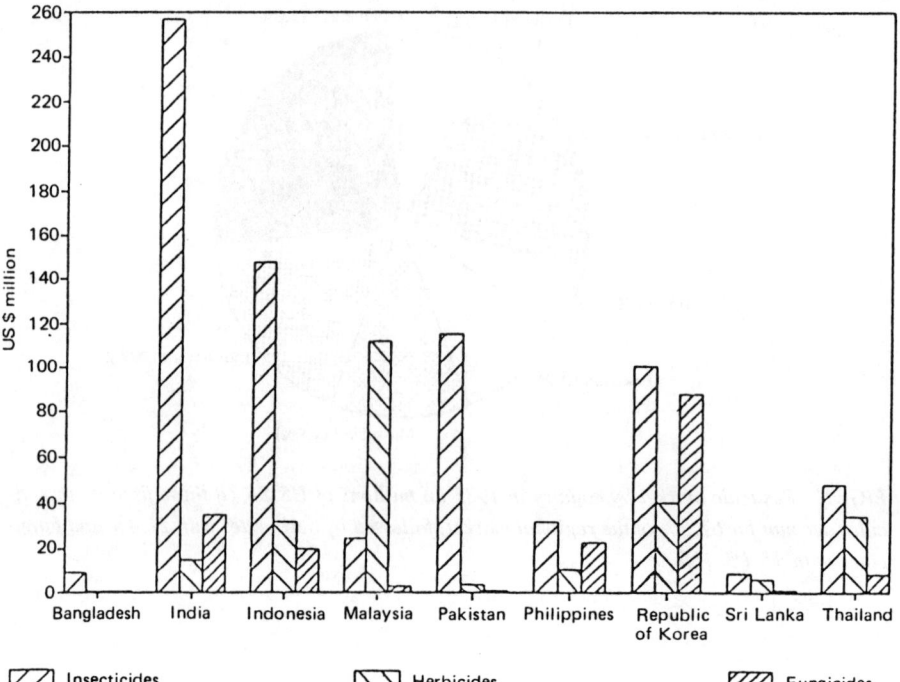

FIG. 2. *Pesticide usage by major types in different countries in Asia (1985) [1].*

Insecticides account for 62.5% of the regional market, followed by herbicides with 21.4% and fungicides with 15.1%. The bulk of the insecticides were used on rice, cotton and vegetables; herbicides are mainly used on rubber, tea, oilpalm, sugarcane and rice; and fungicides are mainly used to control diseases in vegetables and fruits, and to a lesser extent on some plantation crops, e.g. tobacco, rubber and coffee.

The amounts, types and values of the pesticides used in different countries are shown in Fig. 2 [1].

Consumption of pesticides between 1980 and 1985 showed an annual growth of 30% for Indonesia, 20% for Pakistan, 16 to 18% for the Republic of Korea, 12 to 16% for Sri Lanka, 13.6% for the Philippines, 5 to 6% for India, and a minimal increase for Thailand, Bangladesh and Malaysia.

In 1987 it was estimated that annually at least 2 million tonnes of pesticide products are scattered over the global environment. The price is estimated to be US $22.22 billion[1], of which approximately 22% (or US $4.45 billion) were used in developing countries, mainly for cash crops such as cotton, tea, rubber, jute, oilpalm, vegetables and fruits [2].

---

[1] 1 billion = 10^9.

2.3. Pesticide industry

In all Asian countries the pesticide industry is in the hands of the private sector, i.e. a free market situation prevails. However, in some countries the control of government agencies is rather strong, e.g. in Indonesia, particularly on matters relating to regulations and registration. For example, in Indonesia 75% of the pesticides are distributed under government subsidy programmes for mass guidance and mass intensification [1]. Private companies supply all the requirements through government tenders and compete heavily for the remaining 25% of the market [1].

The nine countries under review have pesticide formulation plants that are operated by the private sector and produce finished products. India, Indonesia and the Republic of Korea manufacture a number of technical materials, but in general the region still imports the bulk of its pesticides. Table I shows the local production capacity of pesticides in the nine selected countries [1].

In 1985 it was estimated that about 760 000 tonnes of pesticide formulations were locally produced; this represents approximately 70% of the total formulations used in the nine countries. The balance was imported as finished products. Table II shows the registered pesticides in each of the nine Asian countries [1].

3. ENVIRONMENTAL PROBLEMS OF PESTICIDES

The problems of pesticide usage in Asia fall into four main categories: the pesticide treadmill, pesticide side effects, lack of information and legislative problems [6].

3.1. Pesticide treadmill

Farmers and plantation owners in developing countries are anxious to extract the most from their land and from technological inputs such as pesticides in order to assure a better way of life. However, as this trend will probably result in numerous side effects, in the long term it is unlikely that any production increase will really fulfil the development efforts being made. Thus, the poor will not only have an unequal share of the benefits, but they are also likely to suffer most from the side effects. The situation may become even more serious if in turn a production increase results in more serious pest problems. Killing the natural enemies of a pest of minor importance may cause this pest to develop into a major threat. For instance, over the last 10 years the brown planthopper has developed from a pest of relatively minor importance into the most damaging rice pest in Asia [6]. This is known to have resulted from changes in agricultural practices, e.g. new rice varieties and pesticide usage;

TABLE I. ANNUAL LOCAL PRODUCTION OF PESTICIDES IN SELECTED COUNTRIES IN THE REGION (*in tonnes*)[a]

Developing countries selected	Active ingredients used		Pesticide formulation		
	Name	Quantity	Rated capacity	Actual production	Local production as per cent of total use
Bangladesh	None	None	11 100	2 500	75
India	DDT	4 500	1 665 000	504 238	98
	Malathion	4 500			
	BHC	3 300			
	Parathion	3 200			
	Endosulfan	1 700			
	Zineb	1 400			
	Cu oxychloride	1 100			
	Dimethoate	900			
	Phosphamidon	800			
	Aluminium phosphide	600			
Indonesia	Diazinon	1 523	91 030	48 000	95
	BPMC	1 362			
	MIPC	375			
Malaysia	Paraquat	6 000	50 000	16 000	45
Pakistan	DDT	1 320	47 000	3 800	30
	BHC	1 320			
Philippines	2,4-D	300	46 100	24 00	45
Republic of Korea	Butachlor	2 400	528 900	124 961	99
	Diazinon	900			
	Paraquat	800			
	Phenthoate	500			
	Demeton-S-methyl	250			
	Parathion	200			
	EPN	200			
	Dichlorvos	300			

TABLE I (cont.)

Developing countries selected	Active ingredients used		Pesticide formulation		
	Name	Quantity	Rated capacity	Actual production	Local production as per cent of total use
Sri Lanka	None	None	8 900	2 574	60
Thailand	Paraquat	4 000	60 260	16 500	40
Total		43 750	2 508 290	762 573	70

[a] The unit originally used was tonnes or kilolitres; the figures are only approximate.
Source: Gaston [1].

the increase in pesticide application has decreased pesticide efficacy, leading to further pest crises. In Malaysia, the diamond back moth, a common pest found in vegetables such as cabbages, is now resistant to at least 12 insecticides.

The worst treadmill is the fatal combination of resistance and mortality of the natural enemies of pests. This leads to increasing pesticide application and decreasing efficacy.

3.2. Side effects of pesticides on the environment

The major environmental concern of the pesticide industry is the production process, e.g. the safety of the work place and the surrounding areas beyond the plant boundaries. Precautions have to be taken into consideration which involve plant design, operating procedures, maintenance, monitoring and, above all, management and operators working in a closely integrated manner.

The tragedy of the Bhopal accident which occurred on 3 December 1984 is a typical example of the industrial risks involved in manufacturing pesticides when there is a deterioration in the safety procedures. An estimated 40 tonnes of highly toxic methyl isocyanate (MIC), an intermediate material used to produce the relatively non-toxic insecticide carbaryl, escaped from a tank into the cool night air and spread over the sleeping city of 800 000 people. As MIC is heavier than air, it remained close to the ground, so killing at least 2000 people and injuring more than

TABLE II. REGISTERED PESTICIDES IN THE ASIAN COUNTRIES SELECTED

	No. by active ingredient	No. by formulations
Bangladesh	49	71
India	127	ND[a]
Indonesia	264	409
Malaysia	140	326
Pakistan	148	314
Philippines	158	328
Republic of Korea	177	290
Sri Lanka	94	405
Thailand	153	214
Total	1310	2357

[a] No data available.
Source: Gaston [1].

200 000. It has been assumed that this disaster was caused by improper production procedures and inadequate safety features [2].

Further potential environmental hazards arising from the use of pesticides are the potential side effects on non-target organisms and pollution of the soil, air and water. Table III shows the impact of pesticide applications on non-target organisms [1].

Soil may be contaminated by the disposal of empty pesticide containers, or by drifts from pesticide applications. The fate of pesticide residues in the soil may also create certain problems. The transformation of phenyl mercury fungicides (which are not highly toxic to humans) into methyl mercury may prove to be a serious hazard to humans, since methyl mercury is highly toxic. Furthermore, contamination of the soil may cause underground stream pollution through leaching of the surface wastes [2].

TABLE III. TYPES OF ENVIRONMENTAL HAZARDS ARISING FROM USE OF PESTICIDES ON NON-TARGET ORGANISMS

Compound or group of compounds	Type of application	Possible environmental effects
Insecticides in general	Vector control foliage sprays	Mortality and population changes in non-target arthropods and vertebrates; development of resistance
DDT	General use	Disturbed reproduction in certain species of birds and fish
Endrin	Wet rice	Fish mortality
Endosulfan	Vector control	Fish mortality
Aldrin, dieldrin	Seed dressing	Mortality in seed eating animals; secondary poisoning and population decline in birds of prey
Herbicides in general	General use	Mortality and population changes in non-target plants and invertebrates; secondary effects on host plant dependent arthropods
Fungicides in general	General use	Disturbance of composition soil microflora
Methyl mercury compounds	Seed dressing	Mortality in seed eating animals; secondary poisoning and population decline in birds of prey
Rodenticides in general	Various baits	Mortality in non-target mammals and birds (including secondary poisoning with some rodenticides)

Source: Gaston [1].

Water may be contaminated by pesticide drifts, in original form or as metabolites. This may be more persistent or toxic than the pesticide itself (DDT to DDE, and malathion to malascon and isomalathion).

Developing countries (including the nine countries surveyed in Asia) use about 15% of the total pesticides of the world, which is approximately 2 million tonnes annually. However, developing countries suffer about half the poisoning cases and nearly three-quarters of the deaths. The World Health Organization estimated that 500 000 people have been poisoned by pesticides, with 92 000 deaths [6]. On the basis of these assumptions, the International Organization of Consumer Unions estimated that in 1986, 375 000 people from developing countries had been poisoned, of whom 10 000 had died. Nevertheless, it is important to realize that in developing countries the situation could easily worsen because of the poor health of the farmers [7]. Poisoning can occur during pesticide storage, mixing, application and reuse of containers, or through accidental ingestion [6, 8]. Today, more than 0.5 kg of pesticides per every man, woman and child is manufactured and applied each year [9].

Pesticides can be taken up by the food chain through illegitimate uses such as use of pesticides to kill fish, contamination in storage or transport, or use of treated seeds as food. Also, pesticide residues may reach agricultural products, e.g. food, tobacco, vegetables, fish and other animal products, as a result of use of pesticides in their production. This could affect the export market of agricultural products from developing countries [6]. The various types of environmental hazards arising from the use of pesticides on non-target organisms are shown in Table III.

3.3. Lack of appropriate information and training

The accessibility of pesticide handlers and users to complete, reliable and usable information is of great importance in facilitating appropriate pesticide management. There are various sources of information, namely extension services, labels and sales promotion. The information is only useful if certain conditions are present, e.g. literacy, availability of protective means, money and the attitude of all concerned. Unfortunately, these conditions do not exist in most developing countries in Asia [6].

Pesticide labels are often not written in the local language or they fail to warn in a sufficiently effective manner against possible hazards or to specify the consequent preventive measures that should be taken. Sometimes information is issued that cannot be followed, e.g. 'keep in a cool place', 'wear full protective clothing', 'in the event of an accident call a physician'. It has also been observed that the label is only read, if at all, after purchase, by which time it is too late to decide what to do other than to use the pesticide, even if the instructions cannot be followed. Therein lies the role of advertising. Instead of providing useful warnings and precautions and recommending appropriate pest management practices, the labels indiscriminately promise a complete solution to pest problems, thus providing users

TABLE IV. PESTICIDES THAT ARE BANNED OR RESTRICTED FOR AGRICULTURAL USE IN THE COUNTRIES SELECTED IN THE REGION[a]

Pesticides	Countries in which restricted	Countries in which banned
Aldicarb	Philippines, Republic of Korea, Sri Lanka, Thailand	—
Aldrin, dieldrin, chlordane, heptachlor	All countries except the Republic of Korea	Republic of Korea
Arsenates (Ca/Pb)	—	India
Arsenites (Cu/sodium)	—	Philippines, Thailand, India
Azinphos methyl/ethyl	—	
Binapacryl	—	India
BHC	Philippines, Sri Lanka	Bangladesh, Thailand, Indonesia, Republic of Korea
Carbaryl + BHC	Philippines, Sri Lanka	Bangladesh
Carbophenothion	—	India
Chlordimeform	—	Pakistan, Thailand
DBCP	—	All nine countries
DDT	—	Bangladesh, Malaysia, Indonesia, Thailand, Philippines, Republic of Korea, Sri Lanka
Dicrotophos	Indonesia	India
Disulfoton	Sri Lanka, Thailand	Bangladesh, India
Endosulfan	Bangladesh, Thailand, Sri Lanka	—
Endrin	—	All nine countries
EPN	—	India, Philippines
Ethoprop	Philippines	—

TABLE IV (cont.)

Pesticides	Countries in which restricted	Countries in which banned
Ethyl parathion	Republic of Korea	Bangladesh, India, Malaysia, Indonesia, Pakistan, Sri Lanka, Philippines
Ethylene dibromide	Philippines	Indonesia
Leptophos	—	All nine countries
Lindane	Thailand	Bangladesh
Methamidophos	Indonesia, Sri Lanka	—
Methomyl	Sri Lanka	Malaysia
Methyl bromide	Indonesia, Philippines, Sri Lanka, Republic of Korea	—
Methyl parathion	—	Bangladesh, Sri Lanka
Monocrotophos	Indonesia, Malaysia, Sri Lanka	—
Mephosfolan	—	India
Nitrofen	—	Philippines
Phenamiphos	Philippines, Sri Lanka	—
Phorate	—	Bangladesh
Toxaphene	—	Indonesia, Thailand, Philippines, India
2,4-D	Sri Lanka	—
2,4,5-T	Philippines	Indonesia, Thailand, Sri Lanka, India
Paraquat	Bangladesh, Indonesia, Philippines, Sri Lanka	—
Mercury fungicides	Sri Lanka	Philippines, Republic of Korea

with incorrect information. This is the area of overlap between lack of appropriate information and violation of the Pesticide Code of Conduct [9] (see also Section 3.4).

Further to this matter is the need for proper training for all concerned. For engineers and other personnel involved on the production site, as well as those persons concerned with the retail of pesticides and extension services, it is essential that knowledge of the technical aspects of pesticides is acquired and that certain matters dealing with environmental philosophy, psychology and ethics are understood. This also applies to scientists involved in pesticide studies, research or innovations. Efforts must be made by these groups to simplify their findings and to introduce them step by step, since only innovations that are simple to understand and implement will be effective in persuading pesticide users in the forseeable future [10]. Further, instruction of trainers at the village level is very important. As Achmadi [7] pointed out, much lower exposure to pesticide will be sustained by careful (trained) workers than by careless (untrained) workers. Some operators spray under adverse environmental conditions, while careful workers may wait for better conditions. This is certainly a consequence of proper training. Another way of promoting the safe handling and application of pesticides is to publish booklets as guidance for the retailers, distributors and shopkeepers, e.g. the Safety Guide published in 1984 by the Agricultural Requisites Scheme for Asia and the Pacific/Economic and Social Commission for Asia and the Pacific [11].

3.4. Regulations and legislation

The impact of various problems of pesticide management could potentially be mitigated and controlled by appropriate regulations and legislation. Only the United States of America has any law regarding the export of hazardous pesticides to developing countries. Table IV lists the pesticides that are banned or restricted for agricultural use in the nine countries selected in Asia [1]. However, some of these chemicals are still produced and illegally exported to developing countries.

---

[a] Banned pesticides include all those in the individual lists of banned products from each of the countries, as well as those pesticides where registrations have been officially withdrawn. Those that have been automatically banned or not allowed for registration due to LD_{50} limits imposed by countries such as Indonesia, Bangladesh and Sri Lanka are not included in this table. Restricted pesticides are those registered only for specific uses and in some countries allowed for use only under limited conditions of sale. The list of restricted pesticides in Sri Lanka includes the following additional products: benfuracarb, carbophenothion, carbosulfan, dazomet, deltametri, demeton-S-methyl, dichloropropane, dimethoate, esbiol, fenvalerate, omethoate, permethrin, thiodicarb, benomyl, captan, captafol, edifenphos, phosethyl aluminium, mancozeb, maneb, metalxyl + mancozeb, PNCB, thiram, thiophanate-methyl, butralin, 2,4-D + piperophos, 2,4-D. It should be noted that many of these are under review.

Source: Gaston [1].

In 1976 it was estimated that 40% of the countries of the world had no specific pesticide legislation. Even those that do often fail to enforce it fully [6]. Thus, in most developing countries these ongoing problems still have virtually no appropriate checks or balances.

In a global context, the Food and Agriculture Organization of the United Nations has developed the International Code of Conduct on the Distribution and Use of Pesticides [12]. This sets forth the responsibilities and establishes the voluntary standard of conduct for all public and private organizations engaged in pesticide matters. The Code covers pesticide management in general, including the testing of pesticides, ways of reducing health hazards, labelling, packaging, storage, disposal, advertising and monitoring. Registration of the pesticides is of great importance in controlling the availability, use and risk in each country.

Despite this ideal approach, certain violations of the Code still occur, e.g. the export to developing countries of pesticides which have been banned or whose use has been severely restricted. Other important aspects are advertisements or labels that violate or infringe the Pesticide Code of Conduct [9]. It is therefore urgent that balanced efforts are made to improve education, provide rewards for achievements, punish violators, and promote law enforcements in order to provide better pest management.

4. RESEARCH NEEDS

Based on the current trends in the demand, supply and consumption of pesticides in Asian countries, there is an urgent need to develop strategies for research and studies, e.g.:

(1) The search must continue for an appropriate IPM, in which pesticides could still be used as an important tool. This should include the development of pesticide efficacy and efficiency, where the high selectivity of pesticides should also be promoted. High selectivity, which only affects target pests, helps to keep the non-pest natural components in the system unharmed, so that the diversity of the entire system is not affected. Efficient use should be made of formulation and application techniques in order to avoid a drift of the pesticide to the environment. Efforts should also be made to minimize use of pesticides by developing alternative ways of controlling pests.

(2) Research should be undertaken to lower the pesticide hazards to humans and other biotic and abiotic components, without affecting the effectiveness of the pesticide. This should include efforts to minimize the accumulation or contamination of pesticides in various resources (e.g. drinking water), or products (e.g. food, tobacco), or food chains (e.g. plankton). The study should also cover the persistence of pesticides and the dispersal and accumulation of pesticides in the environment.

(3) The benefit accrued from the protective coating of pesticides and other slow release formulation techniques may be twofold: efficient use of pesticides and a reduction in the drift was well as the wasted pesticide in the environment. One way of preparing a slow release formulation is to dip the pesticide (mixed with crumb rubber) into an irradiated latex.

(4) The resistance of pest species against pesticides is generally absolute, i.e. there is no effect at extremely high rates of pesticide [13]. As the mode of action of pesticide resistant pests is little, if at all, known, further research is essential if the process is to be controlled. Additional benefits are potential crop plants or livestock that may be induced to acquire resistance against damaging pesticides.

(5) The search for induced mutants of crops or livestock that have pesticide resistance properties should be promoted through use of irradiation techniques.

New pest management techniques that may affect the quality of the environment must have an environmental impact analysis document.

It is concluded that pests are an inescapable component of our man made environment; they may also play a certain natural role that is as yet unknown. Their presence and function have to be managed properly and wisely in order to optimize the benefits accruing from our efforts to improve the quality of human life and to maintain continuous environmental quality.

REFERENCES

[1] GASTON, C.P., Pesticide Usage, Registration and Regulatory Practices Among Selected Countries in Asia, Rep. ESCAP/FAO/UNIDO-FADINAP, Economic and Social Commission for Asia and the Pacific, Bangkok (1986) 22-30.

[2] Environmental Aspects of Pesticide Regulation, Marketing and the ESCAP Region, Committee on Agricultural Development, Economic and Social Commission for Asia and the Pacific, Bangkok (1987) 60.

[3] MATSUSHIMA, S., Rice Cultivation for the Millions: Diagnosis of Rice Cultivation and Techniques of Yield Increase, Japan Scientific Society Press, Tokyo (1980) 159.

[4] KOSTERMANS, A.J.G.H., WIRJAHARDJA, S., DEKKER, R.J., "Weeds: Description, ecology and control", Weeds of Rice in Indonesia (SOERJANI, M., KOSTERMANS, A.J.G.H., TJITROSOEPOMO, G., Eds), Balai Pustaka, Jakarta (1987) 24-268.

[5] PANCHO, J.V., SOERJANI, M., Aquatic Weeds of Southeast Asia: A Systematic Account of Common Southeast Asian Aquatic Weeds, UP Los Baños and BIOTROP, Bogor (1978) 130.

[6] Pesticide Handbook: Profiles for Action, 2nd edn, International Organization of Consumer Unions, Penang (1986) 130.

[7] ACHMADI, U.F., Intersectoral Collaboration for Minimizing Behavioural Exposure to Pesticides: Rationale from a Grassroot Study in Central Javanese Agriculture, Dissertation, Griffith University, Brisbane, 1985, 431 pp.
[8] SIM, F.G., Pesticide Poisoning Report: A Survey of Some Asian Countries, International Organization of Consumer Unions, Penang (1985) 80.
[9] GOLDENMAN, G., RENGAM, S., Problem Pesticides, Pesticide Problems: A Citizen's Action Guide to the International Code of Conduct on the Distribution and Use of Pesticides, International Organization of Consumer Unions, Penang (1987) 94.
[10] SOERJANI, M., "An introduction to the weeds of rice in Indonesia", Weeds of Rice in Indonesia (SOERJANI, M., KOSTERMANS, A.J.G.H., TJITROSOEPOMO, G., Eds), Balai Pustaka, Jakarta (1987) 3.
[11] AGRICULTURAL REQUISITES SCHEME FOR ASIA AND THE PACIFIC/ECONOMIC AND SOCIAL COMMISSION FOR ASIA AND THE PACIFIC, Safe Handling and Application of Agro-Pesticides: A Safety Guide for Pesticide Retailers, Distributors and Shopkeepers, ARSAP/ESCAP, Bangkok (1984) 14.
[12] FOOD AND AGRICULTURE ORGANIZATION OF THE UNITED NATIONS, International Code of Conduct on the Distribution and Use of Pesticides, FAO, Rome (1986) 31.
[13] LEBARON, H.M., GRESSEL, J., Herbicide Resistance in Plants, Wiley, New York (1982) 401 pp.

FATE OF ^{14}C-CARBOFURAN IN MODEL RICE/FISH AND RICE/FISH/*Azolla* ECOSYSTEMS*

Jinhe SUN, Jianying GAN, Yongxi ZHANG
Institute of Nuclear Agricultural Sciences,
Zhejiang Agricultural University,
Hangzhou, Zhejiang,
China

Abstract

FATE OF ^{14}C-CARBOFURAN IN MODEL RICE/FISH AND RICE/FISH/*Azolla* ECOSYSTEMS.
Model rice/fish (RF) and rice/fish/*Azolla filiculoides* (RFA) ecosystems were used to study the distribution and transformation of ^{14}C-carbofuran in components of the two ecosystems. Carbofuran applied to paddy soil transferred to components in the ecosystems, whereas the distribution of ^{14}C-carbofuran residues in components was very different. The ^{14}C-carbofuran residues in water were much higher after surface application in the RF ecosystem in study I (1986) than after mixing into the soil in the RF and RFA ecosystems in study II (1987). The residues decreased slowly in study I, but rapidly in study II. Most of the ^{14}C residues was found in the surface soil. Rice plants accumulated ^{14}C-carbofuran slowly; less ^{14}C was recovered in study II than in study I. In comparing the two ecosystems, there were no significant differences in the behaviour of residues in the soil, plants and fish, but residues in water of the RF ecosystem were always higher than those of the RFA ecosystem; those in fish were a little higher in the RFA than in the RF ecosystem. Most of the ^{14}C residues in the soil, plants and fish were unextractable. The extractable residues in fish remained lower than 0.1 ppm in study II (identification of residues is still in progress). In study I, ^{14}C-carbofuran phenol was the main degraded product in the water and soil, and 3-hydroxy-carbofuran was the predominant metabolite in plants and fish after the first 2 weeks.

1. INTRODUCTION

Carbofuran (2,3-dihydro-2, 2-dimethyl-7-benzofuranyl N-methyl carbamate), a systemic N-methyl carbamate insecticide and nematocide, has a very broad spectrum against many crop pests. Studies have been made of the toxicity, metabolism and degradation of carbofuran in the environment since it was discovered in 1968.

---

* Research carried out with the support of the IAEA under Research Contract No. 4276/RB.

In recent years carbofuran and carbaryl have been used increasingly, since they are the most effective insecticides to control the major rice pests: green leafhoppers, brown planthoppers, stem borers and whorl maggots. In many Asian countries, an ecosystem for the production of rice and fish in paddy fields is common practice, and many people depend on fish as it is their major protein source. Owing to the high toxicity of carbofuran to fish and the special conditions of the ecosystem, the safety of carbofuran in ecosystems and the potentially adverse effects to fish in particular should be thoroughly evaluated. However, to date no data have been published. To obtain a complete picture of the fate of carbofuran in rice/fish and rise/fish/*Azolla filiculoides* ecosystems, dual studies, i.e. study I and study II, were carried out in our laboratory.

2. MATERIALS AND METHODS

2.1. Chemicals

Carbon-14-carbofuran (specific activity: 762.7 MBq/mmol; radiochemical purity: greater than 97%) was kindly supplied by the International Atomic Energy Agency. The reference standards carbofuran (99.8%), 3-hydroxy-carbofuran (96.2%) and 3-keto-carbofuran (99.7%) were donated by the Agricultural Chemical Group, the FMC Corporation and the United States Environmental Protection Agency, respectively. All the solvents used in the studies were of analytical reagent grade. The ^{14}C-carbofuran granules were prepared as follows: carbofuran 3G granules were mixed with ^{14}C-carbofuran using methanol to prepare the ^{14}C-carbofuran granules for application [1].

2.2. Soil

Paddy soil from the Zhejiang Agricultural University farm (organic matter: 3.02%; silt: 39%; sand: 4%; clay: 18%; pH 5.9) was sieved through 10 mm mesh before use.

2.3. Rice plants

Xiushui 48, a late season rice that matures in 110 days, and Erjiufeng, an early season variety, were used in study I and study II, respectively.

2.4. Fish

Grasscarp (*Ctenopharyngodon idella* Vahl.) with a length of 3 to 4 cm and a weight of 1.5 to 3 g, and *Tilapia* (*Tilapia nilotica* L.) with a length of 4 cm and a weight of 4.2 g were used in study I and study II, respectively.

2.5. Model ecosystems

The model ecosystem tank (68 × 95 × 45 cm, or 0.65 m^2) was made of hard glass (5 mm thick), supported in a wooden frame, matted with a thick sponge, and kept in a greenhouse. In both studies two tanks were used as replicates.

2.5.1. Study I (1986)

Air dried paddy soil (114 kg) was placed in each tank. Uncontaminated water (kept 1 week before use) was added to a level of 5 cm above the soil surface. Thirty 'hills' of plants (three seedlings per hill), with a space of 15 × 12 cm between each hill, were planted in each tank. One week later, 40 g of $(NH_4)_2SO_4$ (corresponding to 130 kg N/ha) were added and 30 fish were introduced into each tank. Thirty days after transplantation, ^{14}C-carbofuran granules were spread evenly over the soil at a rate of 0.75 kg active ingredient (a.i.)/ha and 0.336 mCi per tank. The paddy water level was then raised to 7.5 cm and maintained at this level during the test period.

2.5.2. Study II (1987)

Three-quarters of the soil (103.5 kg) was added to each tank and sampling tubes made of mesh stainless steel net were set in the soil, which was then flooded; the remaining one-quarter of the soil (34.5 kg) was mixed with the ^{14}C-carbofuran granules dissolved in water and spread evenly over the soil at a rate of 0.80 kg a.i./ha and 0.45 mCi per tank. The water was raised to the 7.5 cm level, the plants were placed in the tanks the following day and 25 fish were introduced into each ecosystem 5 days after treatment. Feed and *Azolla* were periodically added to the RF and RFA ecosystems, respectively. These ecosystems were managed in the same manner as normal rice fields.

3. SAMPLING AND SAMPLE PREPARATION

3.1. Study I

The sampling schedule is shown in Table I.

(1) *Water*. Water was taken (100 mL; 0.25N HCl) and extracted into three 50 mL volumes of CH_2Cl_2. The extracts were combined and then concentrated under vacuum to 5 mL samples for counting [2].

(2) *Soil*. A glass tube (12 mm in diameter) was inserted into the soil to take the soil sample. The wet soil sample was dried at 80°C. Ten grams of dried soil were

TABLE I. SAMPLING SCHEDULE FOR PADDY WATER, SOIL, PLANTS AND FISH

Day	Water	Soil	Plants	Fish
0	*			
1	*	*	*	*
2	*			*
3	*	*	*	*
4	*			*
5	*			
6	*	*	*	
7	*			*
8	*			
9	*			
10	*	*	*	*
11	*			
12	*			
13	*			
14	*	*	*	*
15	*			
17	*			
21	*	*	*	*
28	*	*	*	*
35	*	*	*	*
52	*	*	*	*
93 (harvest)	*	*	*	*

extracted into 50 mL of the mixed solvent (27 g of NH_4Cl + 30 mL of glacial acetic acid + 970 mL of H_2O) at 60°C for 1 hour, with interval shaking [3]. The sample was then filtered, washed with 30 mL of the same solvent, and the filtrate cooled to room temperature. The filtrate was extracted into CH_2Cl_2. The extracts were then treated in the same way as for water.

(3) *Rice plants*. One hill of rice plants was taken randomly, the roots washed with tap water, and then separated into shoots and roots. The samples were chopped finely, dried at 80°C and milled to powder in a blender. Two grams of roots (shoots) were refluxed in a Soxhlet extractor with 50 mL of 0.25N HCl for 1 hour, with interval shaking [4]. The extracts were filtered and further preparations were as above.

(4) *Fish.* One fish was taken from each tank, rinsed with tap water, chopped into small pieces, blended with 15 mL of 0.25N HCl, and extracted by adding 35 mL of 0.25N HCl for 1 hour [5]. The hydrolysate was cooled, filtered and then rinsed with 50 mL of 0.25 N HCl. The hydrolysate was prepared as above.

(5) *Counting and chromatography.* All countings were done by a liquid scintillation counter (model LKB 1217), using 10 mL of a toluene based cocktail (PPO:POPOP:toluene:2-ethoxy ethanol = 5 g:0.4 g:700 mL:300 mL). One millilitre of each final sample was taken for counting. Certain samples with known reference standards were subjected to thin layer chromatography using silica gel GF254, with ether–hexane (3:1) (vol./vol.) as the liquid phase. The reference standards were visualized by adding a 1% potassium permanganate solution. The developed plates were exposed to X-ray films for 4 weeks. The radioactivity of the spots separated on the plates was counted after scraping them into liquid scintillation vials.

3.2. Study II

(1) *Water.* Sampling and preparation were the same as for study I.

(2) *Soil.* One stainless steel net tube was removed, dried, divided into 0 to 10 and 10 to 20 cm pieces and then prepared as for study I.

(3) *Rice plants and fish.* Sampling and preparation were the same as for study I. At certain sampling points, the fish were divided into muscles, bones, gills, scales and intestines and then prepared for counting.

(4) *Counting and chromatography.* The extracted rice plants, soil and fish were combusted with a biological oxidizer (model OX-400). A toluene based cocktail was used to trap the released CO_2 for counting. Counting and chromatographic separations were done as for study I.

4. RESULTS

Regarding plant growth and fish survival, the model exosystems worked well.

4.1. Study I[1]

The distribution of ^{14}C-carbofuran residues among components of the RF ecosystem is shown in Fig. 1. The residual dynamics in water can be described by the non-linear regression equation calculated from the recovery values of the radioactivity in water

---

[1] The residues described were extractable.

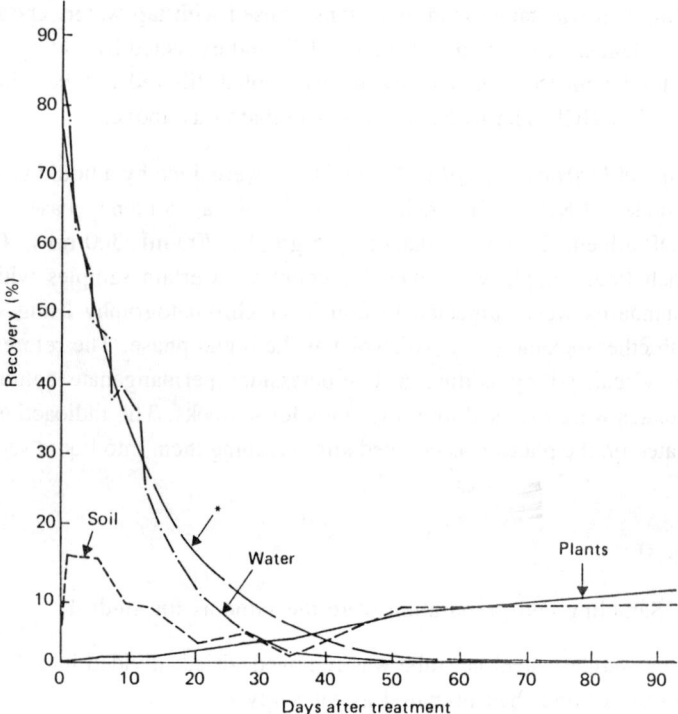

FIG. 1. Distribution of ^{14}C in water, soil and rice plants after surface application of labelled carbofuran (RF ecosystem) (asterisk denotes data from regression equation).

$$C = 55.34e^{-0.091\,t} + 22.29e^{-0.0511\,t}$$

where C is the recovery and t is the time after treatment [6]. The concentration of ^{14}C-carbofuran residues in water decreased to 0.1 ppm (from the initial 0.84 ppm) within 3 weeks.

The residues in the soil reached the maximum level in a very short time, then declined and remained at the same level thereafter, whereas the residues in the rice plants increased with time until harvest. The concentration of ^{14}C-carbofuran residues was always higher in shoots than in roots.

In fish, the ^{14}C-carbofuran residues were very low, and declined to an acceptable limit near harvest.

It was found that carbofuran could be degraded or metabolized in components of the RF ecosystem. Carbofuran phenol was the major and 3-hydroxy-carbofuran and 3-keto-carbofuran were the minor degraded products in water (Table II).

Ten days after treatment, 3-hydroxy-carbofuran was the predominating metabolite in the plants and fish, whereas in the soil, as in water, carbofuran phenol was found in greater quantities (Table III).

TABLE II. CARBOFURAN DEGRADED PRODUCTS MEASURED AS ^{14}C IN WATER (RF ecosystem) (% of total residues in the water)

Degraded products	Days after treatment				
	1	3	6	10	14
Carbofuran	96.2	94.9	91.8	92.4	87.8
3-OH-carbofuran	0.5	1.4	0.7	1.1	1.9
3-O-carbofuran	0.9	0.4	0.5	1.1	1.7
Carbofuran phenol	ND[a]	1.8	6.3	3.7	4.2
Others	2.4	1.5	0.7	1.7	4.5

[a] ND = non-detectable.

TABLE III. CARBOFURAN METABOLITES AS ^{14}C IN SOIL, PLANTS AND FISH (RF ecosystem) (data on day 10; % of total residues in the tank)

Carbofuran metabolites	Soil	Plants		Fish
		Shoots	Roots	
Carbofuran	69.9	69.3	64.0	65.0
3-OH-carbofuran	4.3	18.8	20.5	22.2
3-O-carbofuran	7.8	ND[a]	ND[a]	ND[a]
Carbofuran phenol	8.1	ND[a]	ND[a]	ND[a]
Others	9.9	12.6	15.5	12.8

[a] ND = non-detectable.

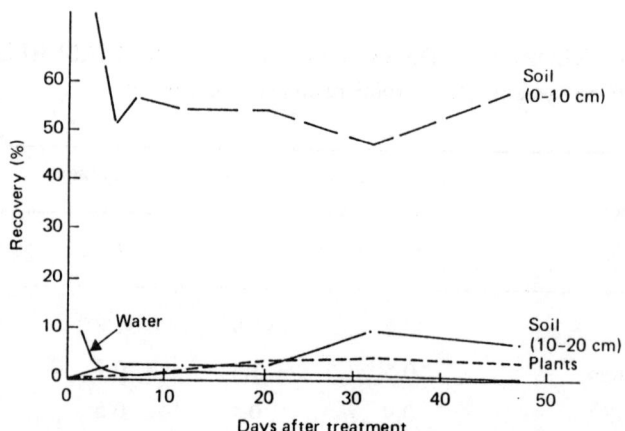

FIG. 2. Distribution of ^{14}C after soil incorporation of labelled carbofuran (RF ecosystem).

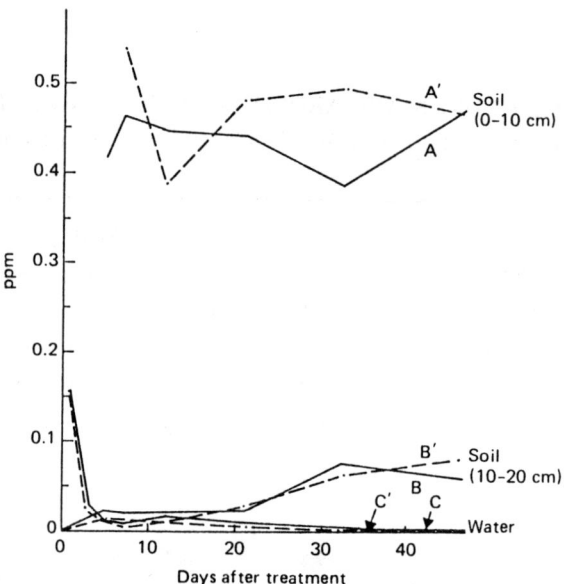

FIG. 3. Distribution of ^{14}C in water and soil after soil incorporation of labelled carbofuran (A, B and C represent the RF ecosystem; A', B' and C' represent the RFA ecosystem).

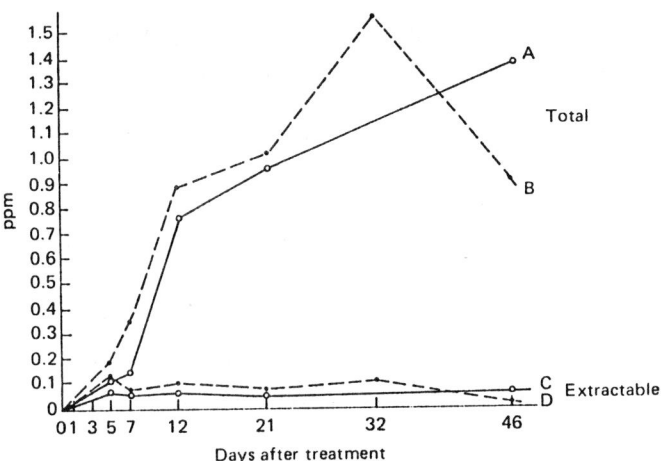

FIG. 4. Distribution of ^{14}C in fish after soil incorporation of labelled carbofuran (A and C represent the RF ecosystem; B and D represent the RFA ecosystem).

4.2. Study II[2]

The carbofuran distribution in the RF ecosystem is shown in Fig. 2. Most of the ^{14}C residues were found in the surface layers of the soil, with little variation. Carbon-14 residues recovered from the lower layers of the soil gradually increased to 7 to 9% of the applied ^{14}C-carbofuran. Obviously it was difficult to eliminate carbofuran from the soil after application of ^{14}C-carbofuran mixed with soil because vertical movement is restricted. The proportion of bound residues in the soil increased rapidly; 92% of the residues were unextractable on day 46. Bound residues play a very important role and the release and bioavailability of carbofuran bound residues warrant further study [7].

In water, ^{14}C-carbofuran decreased rapidly after treatment (Fig. 3). The residue concentration was 0.154 ppm on day 1, which is much lower than that found in study I (0.838 ppm).

Carbon-14 residues in rice plants increased slowly with time (Fig. 2), as also found in study I, and most of the residues recovered in the rice plants were unextractable. However, the ^{14}C recovered was less in study II than in study I. Thus, the method of application may play an important role.

In comparing the RF and RFA ecosystems, no significant differences were found in the behaviour of residues in the soil, water and rice plants. However, ^{14}C residues in water of the RFA ecosystem were always lower than those of the RF ecosystem. This may be attributed to the uptake of *Azolla*.

Considering the relatively small biomass of fish, the ^{14}C recovered never exceeded 0.3% of the total ^{14}C-carbofuran applied. As shown in Fig. 4, most of the

---

[2] Identification of the metabolites is still in progress.

TABLE IV. DISTRIBUTION OF CARBOFURAN RESIDUES AS ^{14}C IN FISH (RF and RFA ecosystems) (% of total residues in the fish)

Ecosystem	% of total residues				
	Gills	Scales	Intestines	Bones	Muscles
RF	6.3	4.1	35.2	32.4	22.0
RFA	3.6	4.1	16.6	35.3	40.4

residues in fish[3] were unextractable, and those that were extractable were always lower than 0.1 ppm; 97% of the ^{14}C residues were bound on day 46 when the concentration was around 1.0 ppm.

Therefore, it is essential that the biological significance of carbofuran bound residues in fish is evaluated. Fish accumulated a slightly higher amount of carbofuran in the RFA than in the RF ecosystem. The distribution of ^{14}C-carbofuran residues in fish is listed in Table IV. Most of the ^{14}C residues were recovered from the intestines, bones and muscles of the fish.

ACKNOWLEDGEMENTS

The authors would like to thank the IAEA for financial support and other assistance, which were extremely valuable in the carrying out of this project, as well as the Agricultural Chemical Group, the FMC Corporation and the United States Environmental Protection Agency for donating the reference standards. They are also grateful to Ziyuan Chen for the interest he showed in this work and to Xingmin Li.

REFERENCES

[1] INTERNATIONAL ATOMIC ENERGY AGENCY, Report on the 1st Research Coordination Meeting on the Use of Isotopes in Studies of Pesticide Residues in the Rice/Fish Ecosystem, Bangkok Jan. 1986 (Internal report).
[2] CARO, J.H., et al., J. Agric. Food Chem. **21** (1973) 1010.

---

[3] Whole fish body.

[3] COOK, R.F., et al., J. Agric. Food Chem. **17** (1969) 277.
[4] BUTLER, L.I., et al., J. Assoc. Off. Anal. Chem. **54** (1971) 1357.
[5] WONG, L., et al., J. Agric. Food Chem. **23** (1975) 315.
[6] GAN, J.Y., CHEN, Z.Y., Acta Sci. Circumst. **6** (1986) 263.
[7] EL ZORGANI, G.A., OMER, I.S., ABDULLAH, A.M., "Bound residues of endosulfan and carbofuran in soil and plant material", Quantification, Nature and Bioavailability of Bound ^{14}C Pesticide Residues in Soil, Plants and Food (Proc. Panel Gainesville, 1985), IAEA, Vienna (1986) 51–56.

IAEA-SM-297/8

STUDIES ON THE EFFECTS OF SOME INSECTICIDES ON THE BRAIN ACETYLCHOLINESTERASE ACTIVITY OF *Tilapia zilli* IN TWO TREATED TROPICAL RIVERS

L.A.K. ANTWI
Institute of Aquatic Biology,
Achimota, Ghana

Abstract

STUDIES ON THE EFFECTS OF SOME INSECTICIDES ON THE BRAIN ACETYL-CHOLINESTERASE ACTIVITY OF *Tilapia zilli* IN TWO TREATED TROPICAL RIVERS.

With a view to controlling onchocerciasis in West Africa, the Marahoue and Black Volta Rivers in Cote d'Ivoire were treated with chlorphoxim and temephos, respectively, at a concentration of 0.5 mg·L^{-1} per 10 minute application to kill the *Simulium* larvae. As part of the Onchocerciasis Control Programme, studies were conducted with caged *Tilapia zilli* to determine the effects of the two larvicides on the fishery resources in the treated rivers. These showed that chlorphoxim inhibits the brain acetylcholinesterase (AChE) activity of the caged *T. zilli* up to 1 km downstream of the breeding site. The highest level of reduction in AChE activity (32%) was recorded in the caged fish placed near the point of release of the chlorphoxim 24 hours after the river treatment. At 0.5 km downstream of the breeding site, the percentage enzyme reduction was 24%, and at the 1 km point the AChE activity was reduced by 17%. There was no significant reduction ($P > 0.05$) in the brain enzyme activity of the caged fish placed at a distance of about 3 km downstream of the breeding site. It was further observed that the caged fish had not recovered from the inhibitory effects of the chlorphoxim 48 hours after the river treatment. No evidence of any inhibitory effects on the brain AChE activity of the caged fish was found as a result of temephos treatment of the Black Volta River at any distance from the point of larvicide application.

1. INTRODUCTION

Onchocerciasis (river blindness) is a filarial parasitic disease which is endemic in the savanna area of the Volta River Basin in West Africa. In this area, which includes parts of Benin, Burkina Faso, Ghana, Cote d'Ivoire, Mali, Niger and Togo (Fig. 1), it is estimated that over 1 million people have the disease, with at least 100 000 who are blind and many more who have impaired vision [1]. The main vector of the disease is the female blackfly, *Simulium damnosum*, which breeds in fast flowing streams and rivers.

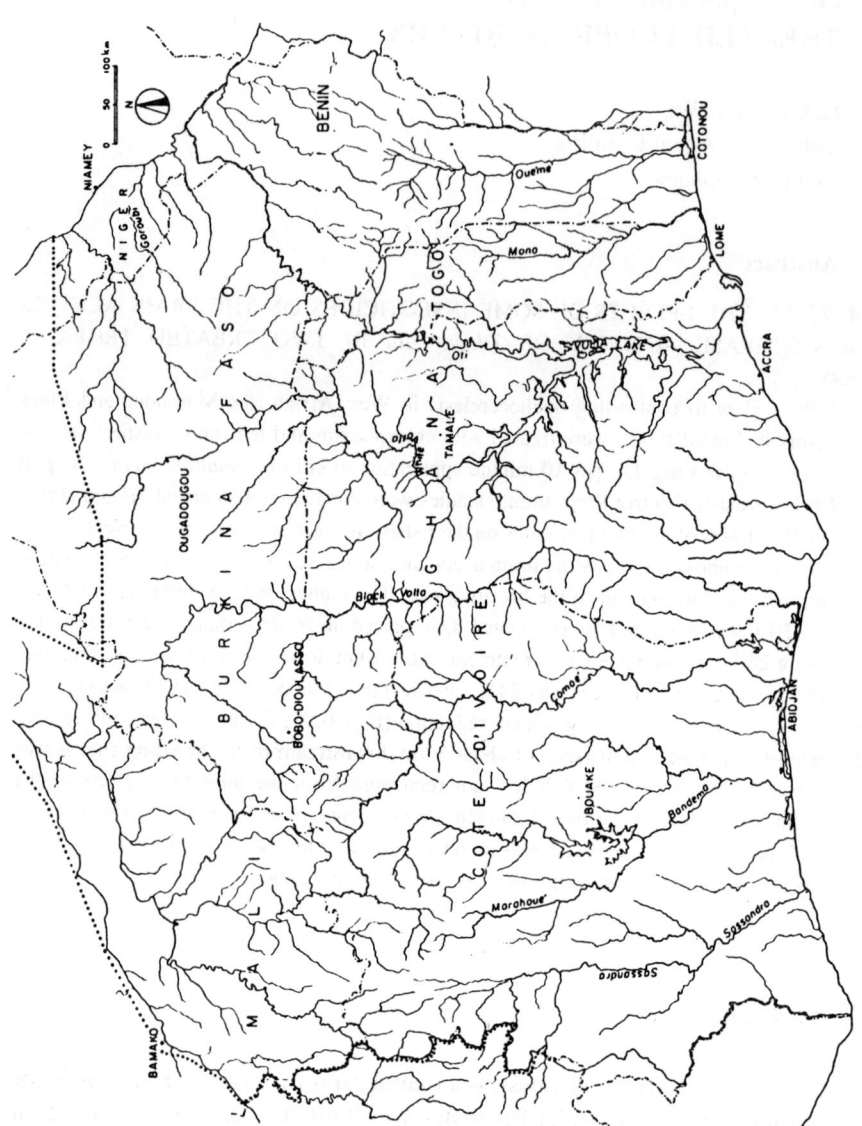

FIG. 1. Onchocerciasis Control Programme in the Volta River Basin area in West Africa.

In 1974 the Onchocerciasis Control Programme (OCP) was launched to control the disease by aerial application of insecticides to all the larval breeding sites of the vector in the programme area. The objective of the OCP was to reduce the risk of blindness in the river basins by lowering the intensity of transmission to a level that does not lead to impaired vision and to maintain this level for at least 20 years [2]. The programme covers an area of 700 000 km^2, of which 14 000 km comprises rivers under treatment [3]. It is supported by the seven West African countries mentioned above, as well as the Food and Agriculture Organization of the United Nations, the International Bank for Reconstruction and Development and the United Nations Development Programme, with the World Health Organization acting as the executing agency.

At the inception of the programme in 1974 the larvicide of choice was the organophosphorus insecticide temephos in a 20% emulsified concentrate (e.c.) formulation. Larviciding was carried out by aerial application of the insecticide to all potential breeding sites of the *Simulium damnosum* at a dosage rate of 0.05 mg·L^{-1} per 10 minute application. Temephos continued to be used in all the programme areas until 1980, when it was observed that two forest cytospecies of the *S. damnosum* complex (*S. sanctipauli* and *S. soubrensi*) had developed resistance to temephos in the rivers in southern Cote d'Ivoire [4]. This unexpected development necessitated the replacement of temephos with a series of chlorphoxim treatments. Nevertheless, temephos in a 20% e.c. formulation continues to be used in most of the OCP areas where the targets are the savanna cytospecies of the *Simulium* complex.

Use of insecticides on a scale as large as that of the OCP requires close monitoring of the effects on non-target organisms and especially on the economically important fishery resources in the treated rivers. One such study carried out on the River Marahoue in Cote d'Ivoire revealed that during larviciding, chlorphoxim inhibited the acetylcholinesterase (AChE) activity of caged fish placed just below the breeding site of *S. damnosum* [5], while temephos treatment of the Black Volta River was found to have no inhibitory effects on the brain enzyme of three species of fish randomly taken from the river [6]. Further to this study, experiments were conducted in the Marahoue and Black Volta Rivers to determine: (1) how far downstream from the point of application is a reduction in the brain AChE activity of fish induced by chlorphoxim during river treatment, and (2) whether temephos inhibits the enzyme activity at any distance away from its point of application. This paper reports on the results of the study.

2. MATERIALS AND METHODS

2.1. Experiment with caged *Tilapia zilli* in the Marahoue and Black Volta Rivers treated with chlorphoxim and temephos, respectively

On the day of the river treatment some *T. zilli* were collected from an untreated barrage with a cast net and transported live to the riverside in plastic bags filled with aerated barrage water. The fish were sorted, divided into groups according to size and then placed in four $52 \times 52 \times 52$ cm cages, each holding about 20 fish ranging from 7 to 13 cm in length. One set of 10 fish was taken for control measurements. The cages were placed in the river 24 hours before treatment at the following distances downstream of the treatment point/breeding site: 0, 0.5, 1.0 and 3.0 km. At about 12:00 hours the larvicide was applied from a helicopter by the OCP team as part of their routine weekly treatment of the river. According to an OCP estimate, the larvicide concentration at the treatment point/breeding site of the blackfly was 0.5 mg·L^{-1} per 10 minute application.

After larviciding, the caged fish remained in the river for 24 hours, after which half of the fish in the cages at each of the four points were removed; each fish was placed in a polythene bag, labelled and placed in an icebox for transportation to the laboratory for enzyme analysis. The rest of the caged fish were removed 48 hours after the river treatment.

2.2. Measurement of acetylcholinesterase activity

The colorimetric method of Ellman et al. [7] was used to measure the acetylcholinesterase activity.

Each fish was decapitated and the operculum removed. The fish head alone was weighed and homogenized using a Potter–Elvehjin homogenizer in a 0.1M phosphate buffer solution (pH of 7.0), so forming a 10% solution. The fish homogenate solution was then diluted with the phosphate buffer to make a 0.1% solution, 4.0 mL of which was measured in a photometric cuvette; 0.1 mL of dithio-bisnitrobenzoic acid (DTNB) solution (0.01M) was added and thoroughly mixed. The spectrophotometric zero was set with this solution, after which 0.1 mL of acetylchiocheline (ASCh) solution (0.2M) was added. Immediately after mixing, the rate of yellow colour production was followed by measuring the absorbance (A) at 412 nm every 30 s for 120 s using a Coleman 295 spectrophotometer. Duplicate measurements were made on each fish homogenate. At the end of each measurement, 4λ anticholinesterase solution was added to check for any non-enzymatic hydrolysis ($\lambda = 0.001$ mL). No increase in the yellow colour was recorded after the addition of the anti-cholinesterase solution, indicating that there was no hydrolysis other than that due to acetylcholinesterase.

The change of absorbance per minute ($\Delta A \cdot \min^{-1}$) was calculated and the rates converted to absolute units using the formula

μmol ASCh$\cdot$min$^{-1}\cdot$g^{-1} fish head weight

$$= \frac{\Delta A \cdot \min^{-1} \times \dfrac{\text{Total volume of solution in cuvette (mL)}}{\text{Volume of homogenate solution taken (mL)}}}{1.36 \times 10^4 \times 10^3}$$

$$\times \frac{100}{\% \text{ homogenate}} \times 10^6 \times \text{head weight (g)}$$

where 1.36×10^4 is the extinction coefficient of DTNB; 10^3 is the conversion of mol$\cdot$L^{-1} to mol$\cdot$mL^{-1}; and 10^6 is the conversion of mol ASCh$\cdot$min$^{-1}\cdot$g^{-1} head weight to μmol ASCh$\cdot$min$^{-1}\cdot$g^{-1} head weight.

The measurements were made at a room temperature of 25 to 30°C.

3. RESULTS AND DISCUSSION

The results of the AChE activity measurements are summarized in Tables I–IV. The enzyme activities are expressed as μmol ASCh hydrolysed per minute per gram fish head weight. Calculation of the percentage inhibition was based on the total enzyme activity of the control fish. The Student's t-test (P = 0.5) was used to compare the enzyme activity of the caged fish with their respective controls.

Table I shows that during larviciding, chlorphoxim at an operational dosage rate of 0.05 mg$\cdot$L^{-1} per 10 minute application inhibits the brain AChE activity of caged *T. zilli* up to about 1.0 km downstream from the breeding site. The highest level of reduction in the AChE activity (31.66%) was recorded in the caged fish placed near the *Simulium* breeding site/treatment point 24 hours after the river treatment. At 0.5 km downstream, the percentage enzyme reduction was 23.72%, and at the 1.0 km point the AChE activity was reduced by 16.53%. There was no significant (P > 0.05) reduction in the brain enzyme activity of the caged fish placed 3 km below the breeding site. During the river treatment, a high concentration of larvicide is released into the river water upstream of the breeding site. As the chemical moves downstream, the large volume of the flowing river rapidly reduces the larvicide concentration to 0.05 mg$\cdot$L^{-1} at the breeding site of the blackfly. At a distance of about 3 km below the breeding site, the larvicide concentration becomes too low to have any significant toxic effect on the fish. Thus, the river water downstream (beyond 3 km) provides a safe place for fish to take refuge during larviciding. It

TABLE I. ACETYLCHOLINESTERASE (AChE) ACTIVITY OF CAGED *T. zilli* RECOVERED FROM THE MARAHOUE RIVER 24 HOURS AFTER CHLORPHOXIM TREATMENT

Distance downstream from breeding site/ treatment point (km)	No. of fish analysed	Weight of fish head (g) (mean ± SD)	Total AChE activity in μmol ASCh·min^{-1}·g^{-1} head weight (mean ± SD)	% AChE inhibition (based on control)
0	10	1.73 ± 0.67	6.28 ± 2.63	31.66
0.5	9	1.75 ± 0.81	7.01 ± 2.81	23.72
1.0	9	1.70 ± 0.52	7.67 ± 2.95	16.53
3.0	8	1.82 ± 0.61	9.80 ± 3.41	0
Control	10	1.79 ± 1.26	9.19 ± 4.15	-

should be mentioned, however, that although the concentration is very low the larvicide may be carried far beyond the 3 km point by the river current.

The significance of AChE inhibition is seen from the point of view of the biochemical importance of the enzyme. AChE controls the passage of impulses in the nervous system of all vertebrates (including fish). Any toxic effect on the enzyme affects transmission of the nerve impulses in the vertebrate and consequently influences its physical and physiological activities. The absence of effective AChE activity in the vertebrate may lead to the loss of muscular co-ordination, convulsion, and ultimately death. No fish died as a result of the chlorphoxim treatment of the River Marahoue. Coppage and Mathews [8] observed that fish died when the brain AChE activity was reduced by 70%, and fresh water fish died when their brain AChE activity was 40 to 70% inhibited [9]. The highest level (32%) of enzyme activity inhibition recorded in this study in the Marahoue River did not cause any fish deaths, but it may result in a reduction in their ability to tolerate reduced oxygen tension [10], and may also affect their feeding activities [11]. Fish under such stress will be prone to easy capture by fishermen and birds, resulting in a greatly reduced population after prolonged chlorphoxim treatment of the rivers.

The results given in Table II show that caged fish had not recovered from the inhibitory effects of the larvicide 48 hours after the river treatment. However, inhibition of the AChE activity of fish is reversible, i.e. fish are able to recover their original enzyme activity level when the source of the water contamination is removed. The recovery rate depends, among other things, on the level of enzyme

TABLE II. ACETYLCHOLINESTERASE (AChE) ACTIVITY OF CAGED
T. zilli RECOVERED FROM THE MARAHOUE RIVER 48 HOURS AFTER
CHLORPHOXIM TREATMENT

Distance downstream from breeding site/ treatment point (km)	No. of fish analysed	Weight of fish head (g) (mean ± SD)	Total AChE activity in μmol ASCh·min^{-1}·g^{-1} head weight (mean ± SD)	% AChE inhibition (based on control)
0	9	1.81 ± 0.52	6.70 ± 3.65	27.09
0.5	8	1.71 ± 0.81	6.87 ± 3.07	25.29
1.0	7	1.72 ± 0.45	7.84 ± 1.19	14.69
3.0	9	1.78 ± 0.71	8.81 ± 3.21	4.20[a]
Control	10	1.79 ± 1.21	9.91 ± 4.15	–

[a] Not significant ($P > 0.05$).

inhibition. Pellissier et al. [12] have observed that in the laboratory it took over 50 days for *Tilapia guineensis* to recover from exposure to 0.05 mg·L^{-1} chlorphoxim for 24 hours when the brain AChE activity was 55.8% of its normal level. Under laboratory conditions experimental fish are under some stress and therefore recovery from such enzyme inhibition takes much longer than in the river water, where they are under no other stress other than the toxic effects of the larvicide. Under natural conditions recovery from the AChE inhibitory effects could be much faster. It is therefore possible that during the OCP operations fish that were exposed to the toxic effects of the larvicide could recover their original AChE activity before the next river treatment.

Tables III and IV show that during treatment of the Black Volta River temephos did not induce any significant ($P > 0.05$) reduction in the AChE activity of caged fish at any distance from its point of release. This result confirms earlier field observations made in similar studies in Burkina Faso [6, 13]. However, in a laboratory study Gras et al. [14] reported that the brain AChE activity of *T. guineensis* was inhibited when they were exposed to 0.05 mg·L^{-1} of temephos for 10 minutes, a phenomenon as yet not observed in the field. A possible explanation could be the physical properties of the temephos formulation used in the OCP operations. The OCP uses a 20% e.c. formulation, which has a density not greater than 0.980 g/mL. It is almost insoluble in water and is easily adsorbed on to surfaces [15]. These properties ensure that during larviciding, especially under non-turbulence conditions, the larvicide remains as a thin film on the surface of the river

TABLE III. ACETYLCHOLINESTERASE (AChE) ACTIVITY OF CAGED T. zilli RECOVERED FROM THE BLACK VOLTA RIVER 24 HOURS AFTER TEMEPHOS TREATMENT

Distance downstream from breeding site/ treatment point (km)	No. of fish analysed	Weight of fish head (g) (mean ± SD)	Total AChE activity in μmol ASCh·min^{-1}·g^{-1} head weight (mean ± SD)	% AChE inhibition (based on control)
0	9	1.94 ± 0.43	8.26 ± 1.36	0
0.5	9	2.09 ± 0.51	7.88 ± 1.49	0
1.0	7	1.80 ± 0.47	7.59 ± 1.09	0
2.0	8	1.65 ± 0.37	7.44 ± 0.79	0
5.0	7	1.86 ± 0.40	8.18 ± 2.14	0
Control	10	1.83 ± 0.39	7.74 ± 2.04	–

TABLE IV. ACETYLCHOLINESTERASE (AChE) ACTIVITY OF CAGED T. zilli RECOVERED FROM THE BLACK VOLTA RIVER 48 HOURS AFTER TEMEPHOS TREATMENT

Distance downstream from breeding site/ treatment point (km)	No. of fish analysed	Weight of fish head (g) (mean ± SD)	Total AChE activity in μmol ASCh·min^{-1}·g^{-1} head weight (mean ± SD)	% AChE inhibition (based on control)
0	8	1.89 ± 0.53	8.28 ± 1.76	0
0.5	7	1.80 ± 0.53	8.21 ± 2.32	0
1.0[a]	–	–	–	–
2.0	6	1.78 ± 0.62	7.81 ± 1.39	0
5.0	8	1.75 ± 0.52	7.84 ± 0.89	0
Control	10	1.83 ± 0.39	7.74 ± 2.33	–

[a] Before larviciding there were only seven surviving fish; these were removed 24 hours after river treatment.

water. Therefore, there is very little mixing of temephos with the whole volume of the river water during treatment, especially in the dry season when river conditions are non-turbulent. (Temephos larviciding is carried out during the dry season, whereas chlorphoxim treatment takes places in the wet season when the river conditions are very turbulent.) Even though the physical properties of the e.c. formulation make temephos larviciding very effective against *Simulim* larvae [15], the chemical has no effect on fish in the treated rivers since very little is absorbed while they are swimming. In addition, other factors contributing to the survival of fish populations in the treated rivers after more than 10 years of weekly temephos treatment include:

(1) The wide safety margin of temephos to fish [16]
(2) The non-persistent nature of temephos, which ensures that the larvicide does not accumulate in the river water after prolonged treatment.

ACKNOWLEDGEMENT

The funds for this study were provided by the Onchocerciasis Control Programme of the World Health Organization.

REFERENCES

[1] Onchocerciasis Control in the Volta River Basin Area, Information Paper, Onchocerciasis Control Programme, Ouagadougou, Burkina Faso.
[2] Onchocerciasis Control in the Volta River Basin Area, UNDP/FAO/IBRD/WHO Joint Report OCP/73.1, Onchocerciasis Control Programme, Ouagadougou, Burkina Faso (1973).
[3] Onchocerciasis Control in the Volta River Basin Area, Annex 111-3, The Control of *S. damnosum* and the Prevention of Environment Contamination, Techniques Criteria for the Selection of Insecticides, Onchocerciasis Control Programme, Ouagadougou, Burkina Faso (1973).
[4] Progress Report of the WHO for 1981, Rep. OCP/PR/81.4, rev. 1, Onchocerciasis Control Programme, Ouagadougou, Burkina Faso (1981).
[5] ANTWI, L.A.K., Environ. Pollut. Ser. A **39** (1985) 151.
[6] ANTWI, L.A.K., Bull. Environ. Contam. Toxicol. **38** (1987) 461.
[7] ELLMAN, G.L., COURTNEY, R.D., ANDRES, V., Jr., FEATHERSTONE, R.M., Biochem. Pharmacol. **7** (1961) 88.
[8] COPPAGE, D.L., MATHEWS, E., Bull. Environ. Contam. Toxicol. **11** (1974) 483.
[9] WEISS, C.M., Ecology **39** 2 (1958).
[10] EATON, J.G., Water Res. **4** (1970) 673.
[11] VERMA, S.R., TYAGYI, A.K., BHATHAGER, M.C., DELALA, R.C., Bull. Environ. Contam. Toxicol. **21** (1979) 502.
[12] PELLISSIER, C., LEUNG TACK, D., GRAS, G., Toxicol. Europ. Res. **IV** 6 (1982).

[13] SCHERINGA, E., STRIK, J.J.T.W.A., ANTWI, L.A.K., Fish Brain AChE Activity after Abate Application against *S. damnosum* in the Volta River Basin, Onchocerciasis Control Programme, Ouagadougou, Burkina Faso, 1981 (unpublished report).
[14] GRAS, G., PELLISSIER, C., TACK, D.L., Etude expérimentale de laction du téméphos sur l'activité acétylcholinestérasique du cerveau du *Tilapia guineensis*, Rep. WHO/UBC/82, World Health Organization, Geneva (1982) 868.
[15] WALSH, J.F., Bull Ent. Res. **75** (1985) 549.
[16] von WINDEGUTH, D.L., PATTERSON, R.S., Mosq. News **26** (1966) 377.

DISTRIBUTION AND METABOLISM OF CARBOFURAN IN PADDY RICE FROM CONTROLLED RELEASE FORMULATIONS*

L. VOLLNER, R. KUTSCHER
FAO/IAEA Agrochemicals and Residues Unit,
IAEA Seibersdorf Laboratory,
Vienna

J. DOMBOVÁRI, M. ONCSIK
Research Institute for Irrigation,
Szarvas, Hungary

Abstract

DISTRIBUTION AND METABOLISM OF CARBOFURAN IN PADDY RICE FROM CONTROLLED RELEASE FORMULATIONS.

This study was conducted to evaluate slow release formulations of carbofuran in aquatic systems. Laboratory, greenhouse and field experiments in a rice paddy were carried out to determine the behaviour and fate of the important systemic insecticide carbofuran after application of new formulations. The release rates into water and the residues in the plant and soil were determined. Metabolites were identified by gas chromatography–mass spectrometry. Carbon-14 labelled carbofuran was used for the experiments. The results showed that the cellulose based (hydroxyethyl-cellulose) formulations released the active ingredient carbofuran at a significantly slower rate than commercial formulations, but that the concentration of carbofuran in water and plants was high enough to protect the plants against insects. Recovery of the insecticide in the rice paddy was only about 30%. Most was lost by changing the water in the paddy field, according to the usual cultivation practice, and by possible evaporation. The radioactivity found in grain and straw was 0.03% and 0.2%, respectively. The distribution of carbofuran in soil at different layers was: 0–5 cm: 17%; 5–15 cm: 7%; 15–25 cm: 4%; 25–40 cm: 2%. Less than 10% of the total radioactivity could be extracted from plants (Soxhlet extraction); thus, most of the carbofuran was metabolized and bound to the plant-grain constituents.

1. INTRODUCTION

Carbofuran (2,3-dihydro-2,2-dimethyl-7-benzofuranyl-N-methylcarbamate) is an important broad spectrum N-methyl-carbamate insecticide, acaricide and nematicide which is effective through contact, stomach and systemic action [1]. It

---

* Research carried out with the support of the IAEA under Research Contract No. 3795/GS.

is applied to foliage at 0.25 to 1.0 kg active ingredient (a.i)/ha for the control of insects and mites, or applied to the seed furrow at 0.5 to 4.0 kg a.i./ha for the control of soil dwelling and foliar feeding insects, or broadcast at 6 to 10 kg a.i./ha for the control of nematodes. The half-life of the compound in a model ecosystem was about 20 days [2], but within plants the carbamate is less persistent. On the other hand, metabolites remain in crops as conjugates and might affect humans directly or at the end of food chains. Also, caution is required because of the potential hazards to humans, from ingestion or inhalation, and also to game birds and animals.

To minimize the hazards to users and to reduce the residues but simultaneously to retain the effectivity of the biological active compound, we formulated carbofuran with different materials, mainly for use in aquatic systems. We selected natural polymers such as latex, sodium alginate and cellulose to avoid any additional negative effects on the environment through toxic or non-biodegradable materials.

Our trials showed that hydroxyethyl-cellulose (Natrosol) [3] cross-linked with resin (melamine-formaldehyde) is a promising agent for the required formulations. We tested this material in laboratory, greenhouse and field experiments and found that its properties were better than those of the commercial formulations.

2. MATERIALS AND METHODS

2.1. Preparation of formulations

Natrosol is a non-ionic, water soluble polymer hydroxyethyl ether derived from cellulose. Natrosol dissolves readily in water and is used to produce solutions with a wide range of viscosity. We used three different types: HHR, HR and GR [3]. Natrosol films can be made water insoluble by reacting them with resin and catalyst. For our experiments we used the following procedure: 0.5 g of NH_4Cl was dissolved in 50 mL of water and heated to 50°C; 0.5 g of Natrosol was then added slowly and mixed. After complete solution, pesticides were added and polymerization was initiated by adding 1.8 mL of melamine-formaldehyde 56% (Reichhold Chemie, Vienna, Austria).

Alginate is a linear polysaccharide derived from brown seaweed. Sodium alginate, the most common algin, is water soluble, with unique thickening, suspending, emulsifying, stabilizing, water holding and gel forming properties. The desired gel is obtained by dissolving 0.5 to 1.5% sodium alginate in water. Pesticide was incorporated into this gel by dissolving it (at a rate of 1 to 20% dry weight) in small amounts of organic solvent (acetone, alcohol), using an Ultra Turrax homogenizer. Commercially available formulations such as emulsifiable concentrate, wettable powder or dispersion can also be used for this purpose; in these cases additional organic solvents are not required.

This formulation is suitable for application as a spray or for the preparation of water insoluble materials. For example, water insoluble granules were prepared by dropping mixtures of alginate gels into $CaCl_2$ solutions (0.25M). After 3 to 5 minutes polymerization occurred and beads could be removed from the media. The granules were then dried at ambient temperature for 2 to 3 days. In a similar way, latex–alginate mixtures were dropped into $CaCl_2$ solutions, so obtaining combined formulations of alginate and latex.

For latex formulations, pesticide was homogenized by stirring with a latex alkaline suspension and dried in Petri dishes at ambient temperature. The alkaline solution of latex can also be dropped into weak acid solutions to coagulate the rubber. In some cases, we also irradiated latex in a gamma source to cross-link the rubber. The effects of the alkaline media (pH 8 to 9) and the gamma ray radiation on carbofuran were checked by high pressure liquid chromatography (HPLC). During the short treatments no significant hydrolysis or degradation occurred.

Sodium alginate of analytical grade was obtained from Fluka AG, Buchs, Switzerland. Natrosol [3] and latex were obtained from Neuber Chemie, Vienna, Austria. The bioactive material carbofuran was obtained as a 75% emulsifiable concentrate (e.c.) formulation (Curaterr) from Bayer AG, Leverkusen, Federal Republic of Germany; the analytical grade came from Supelco, Bellefonte, Pennsylvania, United States of America. Furadan granules with an a.i. of 5% were obtained from FMC Corporation, Princeton, New Jersey, United States of America. Carbon-14 labelled carbofuran (furan ring labelled in positions 2 and 3) with a specific activity of 470 MBq/mmol was supplied by the Hungarian Academy of Sciences, Budapest, Hungary.

For the release experiments, the ^{14}C labelled formulations were submerged in 500 mL of demineralized water at 25°C. Water samples were taken from each container at various times throughout the experiment in order to check the release rates by HPLC and liquid scintillation counting. At the end of the experiment, the formulation matrix was analysed for residues by sample oxidation. For simulation of flowing water, water was slowly stirred at 25°C using magnetic stirrers.

The leaching characteristics of the formulations were evaluated in soil columns. The experimental soil was packed in a glass cylinder with a diameter of 7.5 cm and a height of 8 cm. Unformulated carbofuran was used as a comparison standard.

2.2. Field experiments

The experiments were set in the experimental rice field of the Research Institute for Irrigation, Szarvas, Hungary. The soil was lime conditioned, solonetz type meadow soil. The chemical data of the soil are shown in Table I.

In the experimental field rice had been produced in the previous year. PK fertilizer was sprayed on to the field in September 1985, then it was ploughed

TABLE I. SOME SOIL CHARACTERISTICS OF THE EXPERIMENTAL RICE FIELD

Depth (cm)	pH K$_2$O	pH KCl	Soluble salt (%)	Clay content (<0.001 mm) (%)	Total humus (%)	Total N (%)	Available (Al-P) P$_2$O$_5$ (ppm)	Available (Al-P) K$_2$O (ppm)	CaCO$_3$ (%)
0–20	7.3	6.7	0.008	36.34	2.41	0.13	8	11	0.80
20–40	7.4	6.7	0.09	47.14	2.12	0.10	8	12	0.00
40–60	7.5	6.8	0.09	46.12	1.26	0.09	6	9	0.00
60–80	7.9	6.10	0.10	45.25	1.06	0.06	4	6	2.30

in. In April 1986 the field was harrowed, rolled and then N fertilizer was worked into the soil with a cultivator. Finally, before sowing, the experimental field was compacted with a roller.

The rice was sown on 5 May 1986 at a depth of 30 mm and with a row space of 130 mm. The sowing material was elite, with a germination capacity of 94%. The variety of rice used was Nucleoriza (crossed in Hungary), grown in the largest plot. The growing season was 120 to 125 days.

In a parallel experiment, the commercial formulation Furadan labelled with ^{14}C-carbofuran was applied and compared with the controlled release experiment. The experiment was set in the rice field, in a 1 m^2 metal frame which was placed in undisturbed soil. The height of the metal frame was 40 cm. Carbofuran products were placed next to the roots in the soil. The total activity of carbofuran was 13.5 MBq/m^2 for each plot. After application of the formulations, the plants were flooded. A water depth of 15 cm was retained during the growing season. After 1, 3, 5, 7 and 10 days, later every 10 days, leaves and water samples were collected and the radioactivity measured.

The rice was harvested on 24 September 1986. Grain and straw samples were collected from each of the treatments for radioactivity measurements. Soil samples were also collected from soil layers of 0–5, 5–15, 15–25 and 25–40 cm.

3. RESULTS AND DISCUSSION

Since initial formulations with alginate did not significantly improve the formulation properties, we changed the formulating agents until we finally obtained better results. A Ba-alginate cross-linked matrix with slightly improved properties to those of Ca-alginate and Furadan (see Fig. 1) but even slower release rates into water could be obtained with latex and Natrosol. Latex irradiation experiments for catalysing cross-linkage were also not successful. Finally, we adapted Natrosol, cross-linked with melamine-formaldehyde resin [3], and obtained good results. Figure 2 shows the release of carbofuran from Natrosol, compared with other slow release formulations tested in the laboratory.

For the field experiments we selected Natrosol, since it is cheap, the preparation of formulations is simple and the materials are non-toxic (the formaldehyde content in a water solution is less than 1%).

A comparison of the results obtained with Natrosol and Furadan shows that a few days after application the carbofuran concentration decreased in both cases (Fig. 3). This is obviously due to adsorption in the soil and to uptake of a.i. by rice plants. After another 4 to 5 days, the concentration again increased in both cases; it was significantly higher with our formulation. However, laboratory experiments showed that Furadan has no retaining properties. Therefore, this increase and further residual behaviour can only be explained by evaporation of the water, and release

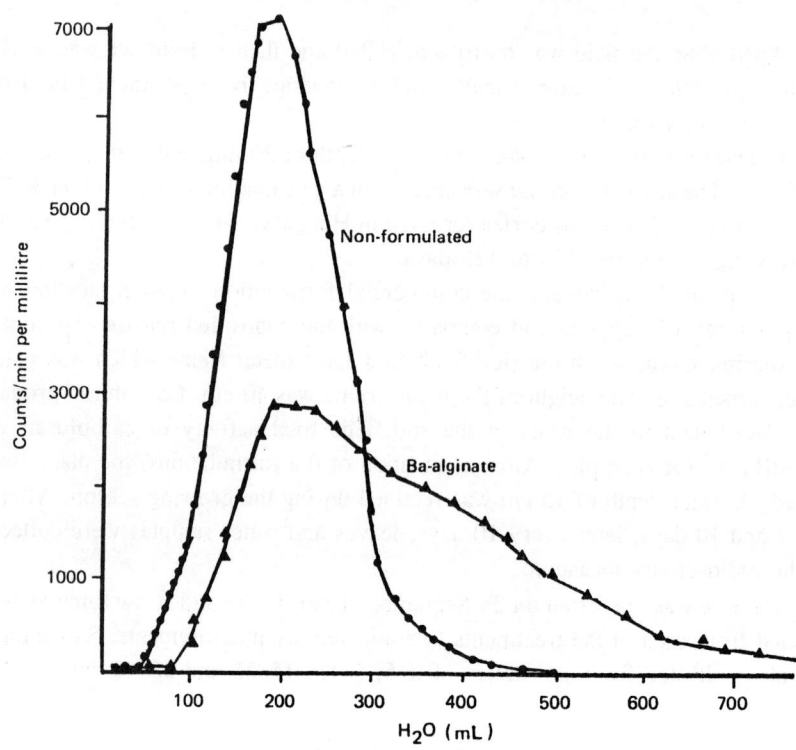

FIG. 1. Leaching of carbofuran from the soil column.

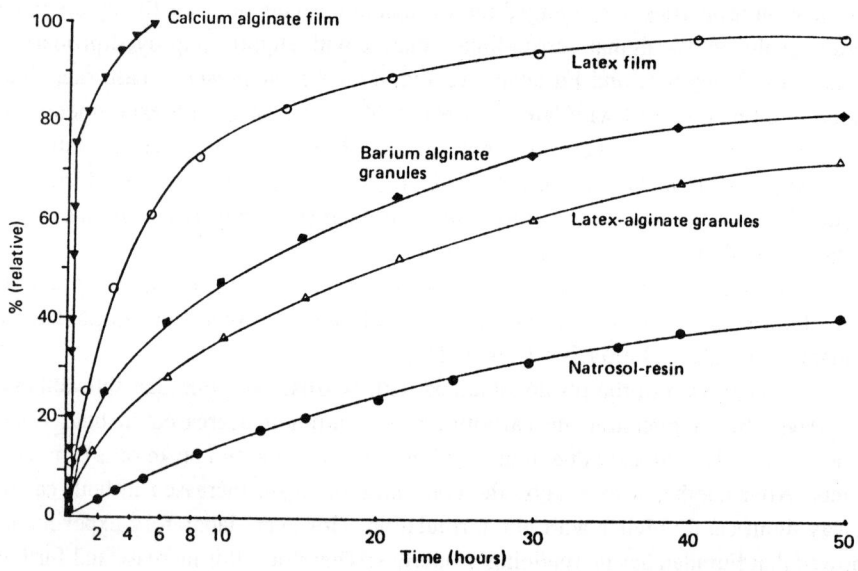

FIG. 2. Relative release rates of carbofuran from different controlled release formulations.

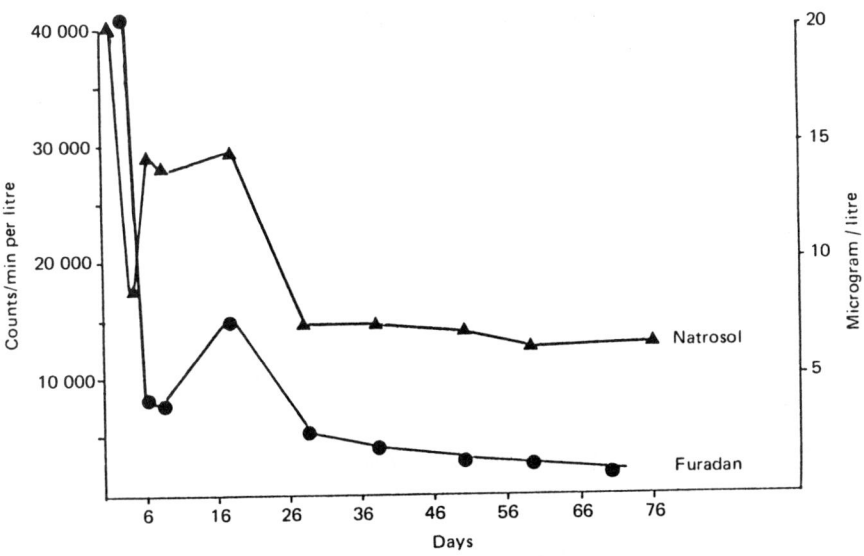

FIG. 3. Concentration of carbofuran in paddy water after application of different formulations.

from the soil. Furthermore, it has been reported that (non-protected) carbofuran rapidly metabolizes in paddy water [2]. These effects can be avoided by using the Natrosol formulation.

Although the a.i. concentration levels from controlled release formulations are higher in water and plants throughout the whole vegetation period (Fig. 4), residues in the grain and straw are not higher than those using Furadan (0.3 to 0.7 ppm). Tables II–IV show the detailed residue analysis results.

Collected extracts of plant and grain material were analysed by gas chromatography–mass spectrometry after cleanup on gel columns. Figure 5(a) shows a total ion chromatogram of such an extract; Fig. 5(b) shows combined ion chromatograms of residues and metabolites. The characteristic ion 164 of carbofuran appears after the breakdown of methylisocyanate (MW 221-57). The ion chromatogram of the analytical standard of carbofuran also indicates an isomer as by-product. Both compounds could also be detected in plant and grain samples (15 and 8%, respectively). After methylation of the extract we detected 3-methoxy-carbofuran, which corresponds to 3-hydroxy-carbofuran with the ion 193 (MW 251-58) (30%), 3-methoxy-carbofuran-(7)-phenol-CH_3, which corresponds to 3-hydroxy-carbofuran-(7)-phenol with the molecule ion 208 (5%), carbofuran-(7)-phenol-CH_3, which corresponds to carbofuran-(7)-phenol with the ion 178 (20%), and 3-ketocarbofuran-(7)-phenol-CH_3, which corresponds to 3-ketocarbofuran-(7)-phenol with the ion 162 (MW 192-31+1) (21%). No ketocarbofuran could be found [4].

TABLE II. FINAL RESIDUES OF CARBOFURAN AND METABOLITES (in mg). TOTAL APPLIED: 360 mg

Formulation	GR-Natrosol	HR-Natrosol	HHR-Natrosol	Furadan
Rice grain	0.2	0.1	0.1	0.1
Rice straw	0.7	0.6	0.5	0.4
Soil 0– 5 cm	(40.8)	77.8	64.1	31.6
Soil 5–15 cm	27.7	31.7	24.5	10.7
Soil 15–25 cm	8.7	10.8	16.8	2.3
Soil 25–40 cm	(26.7)	9.6	8.2	4.2

Note: Numerals in brackets indicate that some irregular leaching has occurred.

TABLE III. FINAL RESIDUES (in %). TOTAL APPLIED: 100%

Formulation	GR-Natrosol	HR-Natrosol	HHR-Natrosol	Furadan
Rice grain	0.05	0.03	0.03	0.02
Rice straw	0.21	0.18	0.14	0.11
Soil 0– 5 cm	(11.3)	21.6	17.8	8.8
Soil 5–15 cm	7.7	8.8	6.8	3.0
Soil 15–25 cm	2.4	3.0	4.7	0.6
Soil 25–40 cm	(7.4)	2.6	2.3	1.2

Note: Numerals in brackets indicate that some irregular leaching has occurred.

TABLE IV. RESIDUES (in relative percentage). TOTAL FOUND: 100%

Formulation	GR-Natrosol	HR-Natrosol	HHR-Natrosol	Furadan
Rice grain	0.16	0.08	0.09	0.09
Rice straw	0.71	0.48	0.44	0.43
Soil 0– 5 cm	38.9	59.6	56.1	27.7
Soil 5–15 cm	26.5	24.3	21.5	9.4
Soil 15–25 cm	8.3	8.3	14.7	2.0
Soil 25–40 cm	25.4	7.3	7.2	3.8

IAEA-SM-297/28 265

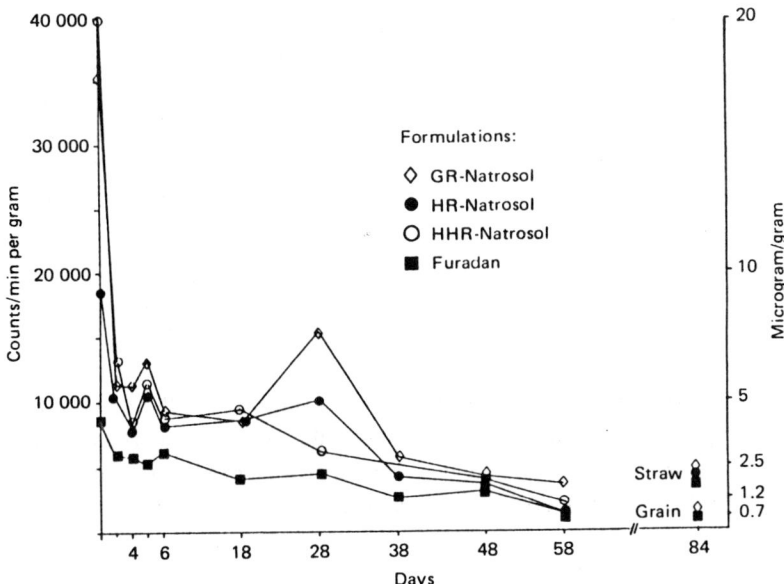

FIG. 4. Carbofuran residues in rice plants during the vegetation period.

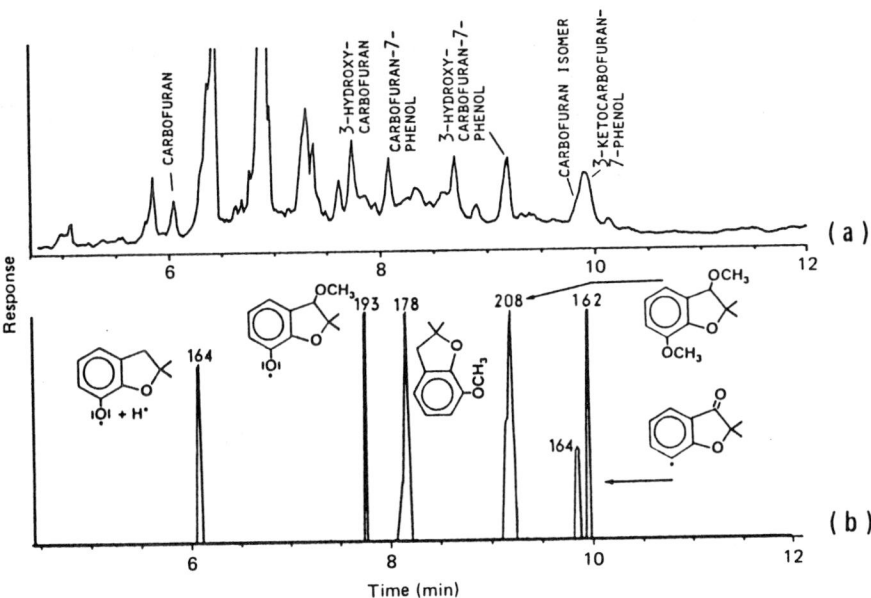

FIG. 5. (a) Total ion chromatogram of rice extract, and (b) combined ion chromatograms for identification of metabolites after methylation.

It was concluded that by using different natural polymers, and some combinations of them, release rates of carbofuran in aquatic systems could be slowed down. Simple formulation procedures and low prices would also permit application of formulations in less developed countries. Laboratory and field experiments showed improved environmental behaviour compared with Furadan, e.g less runoff and less evaporation. Although slightly higher residue levels in paddy soil were observed, no increase of residues in rice grain was found.

ACKNOWLEDGEMENTS

The authors gratefully acknowledge the Government of the Federal Republic of Germany for supporting the project and the technical assistance given by L. Szilvassy, Szarvas, Hungary, and C. Pascucci, FAO/IAEA Agrochemicals and Residues Unit, IAEA Seibersdorf Laboratory, Vienna.

REFERENCES

[1] The Pesticide Manual, 8th edn, The British Crop Protection Council, Thornton Heath, Surrey (1987).

[2] YU, C.C., BOOTH, G.M., HANSEN, D.J., LARSEN, J.R., J. Agric. Food Chem. **18** (1970) 92.

[3] NATROSOL, Hydroxyethyl-cellulose — Physical and Chemical Properties, Hercules BV, The Hague, Netherlands.

[4] ARCHER, T.E., STOKES, J.D., BRINGHURST, R.S., J. Agric. Food Chem. **25** (1977) 536.

IAEA-SM-297/33

IMPROVED ALGINATE BASED SLOW RELEASE PESTICIDE FORMULATIONS*

E. SCHACHT, J.C. VANDICHEL
Laboratory of Organic Chemistry,
State University Gent,
Gent, Belgium

Abstract

IMPROVED ALGINATE BASED SLOW RELEASE PESTICIDE FORMULATIONS.
Use of poly(ethylene imine) treated Ca-alginates was investigated for the preparation of slow release formulations of dichlobenil, propanil and carbofuran. It was demonstrated that release of pesticides from the alginate granules was markedly retarded by post-treatment of the Ca-alginate beads with polyamine. The release profile depended on the post-treatment procedure: type and concentration of the polyamine, pH and duration. Retardation of herbicide release up to 1 year was attained.

1. INTRODUCTION

Current interest is being shown in developing slow release formulations of pesticides. The major objective is to achieve more efficient use of the biocide, so avoiding the need for repeated applications or use of large doses. Consequently, the environmental impact is more favourable.

Most slow release formulations are based on a physical combination of the pesticide with a polymer. Among the large number of polymer materials that have been selected for this purpose are alginates. Alginic acid is a natural polysaccharide composed of 1,4 linked α,L-guluronic acid and β,D-mannuronic acid units. In the sodium salt form the polymer is water soluble. Addition of a bivalent cation, e.g. Ca^{2+} salts, causes gelation of the polysaccharide due to ionic cross-linkage of the individual polymer chains [1].

Pesticide formulations based on Ca-alginate gels have been described in the literature [2-6]. When placed in water Ca-alginate gels release the enclosed active agent at a rate which is dependent on the water solubility of the agent [6].

It has been reported [7-9] that treatment of Ca-alginate gels with cationic polymers, e.g. polyamines such as poly(ethylene imine) (PEI) or polylysine, leads to the formation of surface coated gels which have superior stability properties. This approach has been successfully applied for the immobilization of viable cells.

---

* Research carried out in association with the IAEA under Research Agreement No. 3470/CF.

We investigated the applicability of this post-treatment approach for the preparation of pesticide formulations with improved slow release characteristics. In this paper the production and release properties of PEI treated Ca-alginate beads containing dichlobenil, propanil or carbofuran are discussed.

2. MATERIALS AND METHODS

2.1. Chemicals

Sodium alginate (Satialgine SG 500) was kindly provided by Mero Rousselot Satia Benelux (Brussels, Belgium). The poly(ethylene imines) used were either the BASF products G 20 ($M_n \sim 800$) and Polymin P ($M_n \sim 150\ 000$) (Ludwigshafen, Federal Republic of Germany) or the Cordova Chemical Company product Corcat 12 ($M_n \sim 1200$) (North Muskegeon, Michigan, United States of America). Dichlobenil was obtained from Duphar (Weesp, Netherlands), propanil from Bayer (Leverkusen, FRG) and carbofuran from the FMC Corporation (Philadelphia, USA).

2.2. Production of alginate granules

A typical procedure for the preparation and post-treatment of the Ca-alginate beads is as follows: a suspension of 0.1 to 0.2 g pesticide in 100 mL of a 1% (wt/vol.) aqueous solution of sodium alginate is added dropwise (via a peristaltic pump and through a 3 mm nozzle) to a calcium chloride solution (0.25M). This addition occurs stepwise; after 5 min the addition is interrupted and the gel beads are allowed to cure for another 2 min in the calcium chloride solution. They are then separated by filtration and washed with distilled water. The filtrate is further used for addition of another portion of the sodium alginate–pesticide suspension. The Ca-alginate beads are subsequently transferred to 100 mL of an aqueous PEI solution (10 to 20% (wt/vol.); pH: 1 to 5) and stirred for a given period of time. They are then separated, washed with water to remove excess polymer and finally air dried.

2.3. Release experiments

Release experiments took place in reconstituted fresh water [10] under static conditions at room temperature (25°C). Dried beads (100 mg) were transferred into a 500 mL stoppered Erlenmeyer flask charged with 300 mL of reconstituted water. At regular intervals the water was replaced with fresh medium and analysed for the pesticide by means of high pressure liquid chromatography (HPLC).

2.4. Determination of the pesticide content in the dried beads

A well known amount of air dried beads (approximately 100 mg) was added to an aqueous solution of sodium hexametaphosphate (5% (wt/vol.); pH: 7) and vigorously stirred with a magnetic stirring bar for 6 hours. Then 50 mL of methanol was added and the mixture was centrifuged (20 min, 2000 rev/min). The concentration of pesticide in the supernatant was determined by HPLC.

2.5. HPLC analysis of the pesticides

HPLC analysis was performed on a RSIL C-18 HL column (inner diameter: 0.4 cm; length: 20 cm; 10 μm) (Alltech Europe, Eke, Belgium). As eluent, a methanol–water mixture of the following composition (vol./vol.) was used: 70:30 for dichlobenil and propanil, and 60:40 for carbofuran. Detection was made by UV ($\lambda_{exp.}$ dichlobenil: 237 nm; propanil: 249 nm; carbofuran: 282 nm).

3. DISCUSSION

3.1. Formulation aspects

Upon contact with the calcium chloride solution, the droplets of the sodium alginate solution containing the pesticide gelify immediately, forming highly swollen beads with an average diameter of 4 mm. Upon treatment of these beads with the PEI solution, they shrink to about 25 to 30% of their original diameter, the largest shrinkage being at a pH of 3. It can be anticipated that expulsion of water during this shrinkage process will be accompanied by a significant loss of pesticide. As illustrated in Table I, this loss is higher for the more water soluble pesticides and occurs mainly during the treatment with PEI.

TABLE I. LOSS OF PESTICIDE DURING THE PREPARATION OF THE ALGINATE FORMULATIONS[a]

Pesticide	Water solubility (ppm)	% loss In CaCl$_2$	% loss In PEI	Total loss (%)
Dichlobenil	18			25.5
Propanil	225			42.5
Carbofuran	700	15.2	38.8	54.0

[a] Post-treatment with a 10% solution of Corcat 12, at a pH of 3 for 2 hours.

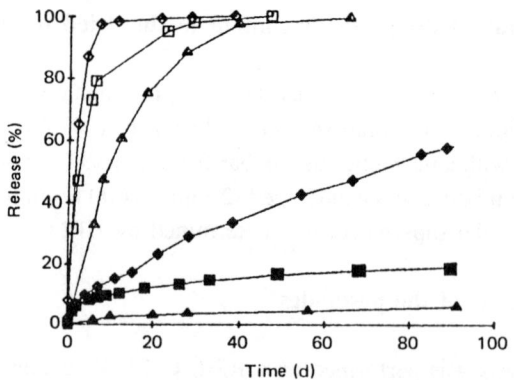

FIG. 1. *Release of dichlobenil (Δ), propanil (□) and carbofuran (◊) from dried Ca-alginate beads (open symbols) and PEI treated alginate (closed symbols) granules (treatment: 10 vol. % Corcat 12; pH: 3; 2 hours).*

3.2. Release experiments

For laboratory evaluation of the formulations, release experiments were carried out under standardized conditions using reconstituted fresh water (pH: 8) as release medium. It was noticed that the dried PEI treated alginate granules swell much less than the dried Ca-alginate beads. Release of the pesticides from the alginate granules varied over a wide range and depended on the type of pesticide and the post-treatment procedure. In general, post-treatment of the Ca-alginate beads resulted in a remarkable decrease of the release rate, as is illustrated in Fig. 1. Under the stated experimental conditions the dried Ca-alginate beads release the active agent within a period of 1 to 6 weeks, whereas for the PEI treated granules a steady release over a period of several months is observed. Release periods exceeding 1 year were observed for some dichlobenil formulations.

For the same post-treatment procedure the release rate increases with the increasing water solubility of the pesticide. This is in good agreement with the results reported recently by Pfister et al. [6] for a series of herbicide containing Ca-alginate beads. It was further observed that for a given pesticide the release profile depends on the type of PEI, the concentration and pH of the PEI solution and the duration of the post-treatment of Ca-alginate beads.

3.2.1. Effect of the pH of the PEI solution

Suhaila and Salleh [9] prepared PEI alginate gels at a pH of 9.6 and investigated the effect of the pH on the shrinkage and rigidity of the obtained beads. It

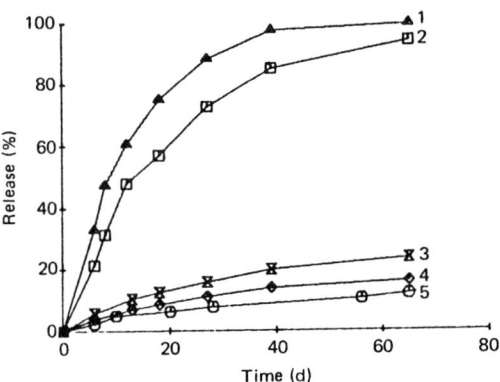

FIG. 2. Effect of the pH of the post-treatment solution on the release rate of dichlobenil from air dried alginate granules (treatment 1: none; 2: 0.1N HCl; 3-5: 10 vol.% Corcat 12; 2 hours; and 3: pH: 5; 4: pH: 1; 5: pH: 3).

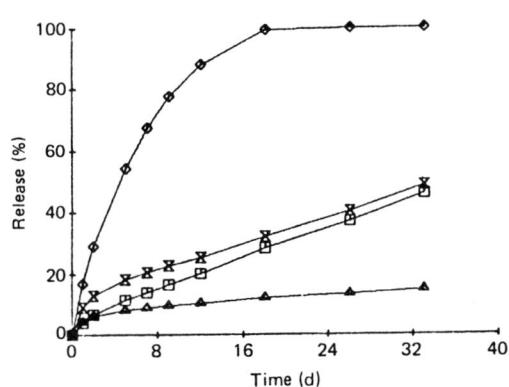

FIG. 3. Release of propanil from alginate granules obtained by treatment of Ca-alginate beads with different PEIs (20 vol.% PEI; pH: 3; 2 hours; ◇: none; □: G 20; ×: Polymin P; △: Corcat 12).

was concluded that in the pH region of 3 to 6 the shrinkage increases with the decreasing pH. Likewise, the rigidity of the gels increased as the pH was lowered. We investigated the effect of the pH of the PEI solution (pH range: 1 to 5) on the release profile. As illustrated in Fig. 2 for dichlobenil alginate formulations, the release rate is minimal for formulations treated at a pH of 3. A similar trend was noticed for the other pesticides.

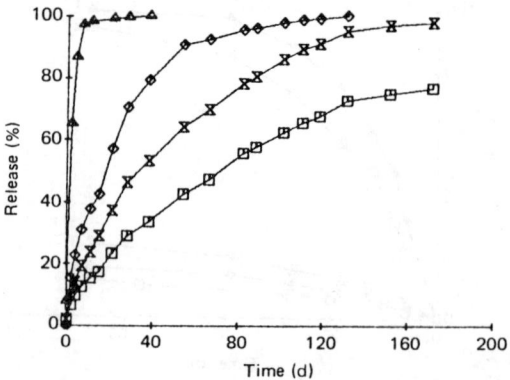

FIG. 4. *Release of carbofuran from alginate granules obtained by treatment of Ca-alginate beads with different PEIs (20 vol.% PEI; pH: 3; 2 hours; △: none; ◇: G 20; ×: Polymin P; □: Corcat 12).*

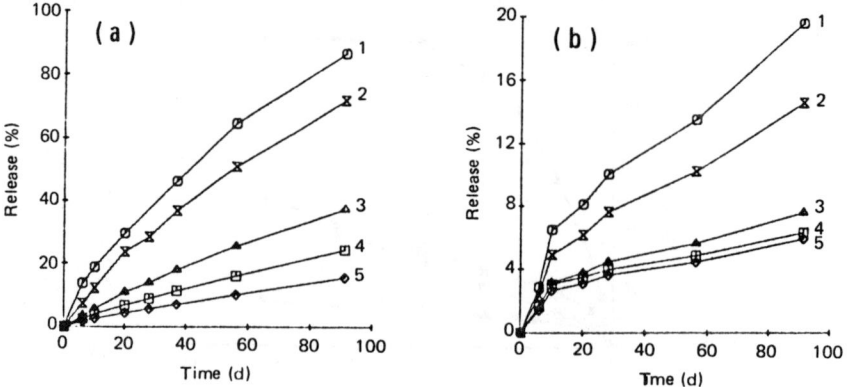

FIG. 5. *Effect of the concentration of PEI and the duration of the post-treatment on the release of dichlobenil from PEI treated alginate granules (treated with Corcat 12; pH: 2; PEI = (a) 5%; (b) 10% (wt/vol.)). 1: 10 min; 2: 20 min; 3: 40 min; 4: 60 min; 5: 120 min.*

Potentiometric titration of PEI demonstrates that maximal protonation, and hence maximal charge density, is obtained at a pH of 3 [11]. At this pH, polyelectrolyte formation may be optimal, which could explain the phenomena observed.

It is expected that in a strong acidic medium the alginic carboxylate groups are protonated, forming more hydrophobic alginic acid. This may explain the slightly retarded release observed after treatment of the Ca-alginate formulation in an aqueous pH of 1. However, this decrease in release rate is inferior to that observed after PEI treatment at the same pH. Clearly, interaction between the polysaccharide and the protonated PEI is the dominating process.

3.2.2. *Effect of the type and concentration of PEI and the duration of the treatment*

Pesticide containing alginate beads were treated with different kinds of PEI (G 20, Polymin P and Corcat 12) at a pH of 3 for 2 hours. As illustrated in Figs 3 and 4 for propanil and carbofuran formulations, respectively, the slowest release is obtained after treatment with Corcat 12.

Since Corcat 12 and G 20 have a molecular weight of the same order of magnitude, the size of the PEI cannot be the determining factor. Commercial PEIs are branched polymers and consequently contain primary as well as secondary and tertiary amino functions

$$\cdots-CH_2CH_2-NH-CH_2CH_2-N-CH_2CH_2-NH-\cdots$$
$$CH_2$$
$$CH_2$$
$$NH$$
$$\vdots$$
$$CH_2$$
$$CH_2$$
$$NH_2$$

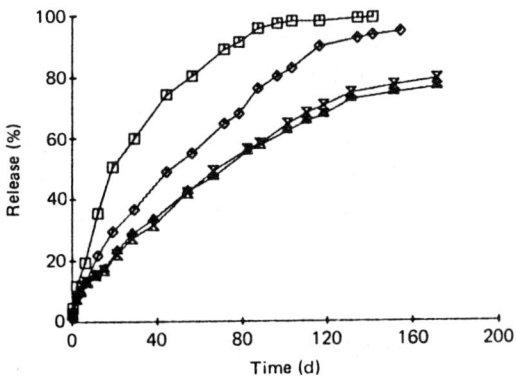

FIG. 6. *Effect of kaolin on the release of carbofuran from PEI treated alginate granules (treated with Corcat 12; 10%; pH: 3; 2 hours); wt% kaolin:* $\triangle = 0$; $\times = 10$; $\diamond = 30$; $\square = 50$.

Titrimetric analysis of the different polyamines revealed that the content of primary amino functions in Corcat 12 is higher (33%) than those of G 20 and Polymin P (24%). Assuming that terminal primary amino functions are more accessible for polyelectrolyte formation with the alginate, these structural differences could help to explain the observed differences in performance among the three polyamines.

The influence of the concentration of the PEI solution and the duration of the post-treatment on the release profile are illustrated in Fig. 5(a) and (b) for dichlobenil formulations treated with Corcat 12 at a pH of 3. Prolonged contact of the Ca-alginate beads with polyamine results in an increased retardation effect. Moreover, increasing the PEI concentration from 5 to 10% results in a remarkable decline in the release rate.

Apparently, increasing the PEI concentration and prolonging the incubation time favours the penetration of PEI in the alginate bead and the interaction with the Ca-alginate. This can lead to a thicker or denser surface coat.

3.2.3. Effect of added kaolin

Kaolin is often added to pesticide formulations to increase the density of the granules or to decrease the price. The effect of kaolin on the release of pesticides from PEI treated alginate granules is demonstrated in Fig. 6. Incorporation of 10% additive does not significantly alter the release profile. However, addition of higher quantities of kaolin leads to a faster rate of release. This is most likely due to the increased heterogeneity and/or the increased porosity of the granules.

4. CONCLUSIONS

Post-treatment of Ca-alginate beads with PEI is an interesting approach to obtain pesticide formulations that can release the active agent over an extended period of time. By proper choice of the type of PEI and the post-treatment procedure, release profiles can be varied over a wide range. This technique is particularly interesting for pesticides with a low water solubility. For good water soluble pesticides the final load of active agent in the granules is low due to considerable losses occurring during the formulation process.

ACKNOWLEDGEMENTS

The authors would like to thank A. Lemahieu from Mero Rousselot Satia Benelux (Brussels) for providing the alginate. This work was supported by the PREST programme of the Belgian Government (Contract No. 12000784).

REFERENCES

[1] GRANT, G.I., MORRIS, E.R., REES, D.A., SMITH, P.G., THOM, D., FEBS Lett. **32** (1973) 195.
[2] BARRETT, P.R.F., Pestic. Sci. **9** (1978) 425.
[3] CONNICK, W.J., BRADOW, J.M., WELLS, W., STEWARD, K.K., VAN, T.K., J. Agric. Food Chem. **32** (1984) 1199.
[4] CONNICK, W.J., J. Appl. Polym. Sci. **27** (1982) 3341.
[5] BAHADIR, M., PFISTER, G., Ecotoxicol. Environ. Saf. **10** (1985) 197.
[6] PFISTER, G., BAHADIR, M., KORTE, F., J. Control. Rel. **3** (1986) 229.
[7] VELIKY, I.A., WILLIAMS, R.E., Biotechnol. Lett. **3** 6 (1981) 275.
[8] LIM, F., U.S. Patent 4.352.883 (1982).
[9] SUHAILA, M., SALLEH, A.B., Biotechnol. Lett. **4** 9 (1982) 611.
[10] STEPHAN, C.E., Rep. 860/3-75-009, United States Environmental Protection Agency, Washington, DC (1975).
[11] BOIS van TRESLONG, C.J., STAVERMAN, A.J., J. Royal Dutch Chem. Soc. **93** 6 (1974) 171.

APPLICATION OF *Bacillus sphaericus* IN THE CONTROL OF *Culex fatigans*

S.V. AMONKAR, A.S. RAO, V. NARAYANAN
Modular Laboratory,
Pest Control Section,
Biochemical Group,
Bhabha Atomic Research Centre,
Bombay, India

Abstract

APPLICATION OF *Bacillus sphaericus* IN THE CONTROL OF *Culex fatigans*.
    A terminal spore bearing bacteria strain (ISPC-5) was isolated from the diseased larvae of *Culex fatigans* and identified as *Bacillus sphaericus* (WHO 2173). It was of the phage type IV and serotype H-26a and 26b. The LC_{50} for *C. fatigans* 2nd instar larvae was 1.7×10^4 colony forming units (CFU)/mL. Comparative toxicity studies made on ISPC-5, as well as on 1593 and *Bacillus thuringiensis* var. *israelensis* (H-14), revealed that this organism was pathogenic to laboratory reared non-resistant and chemical insecticide resistant species of the *Culex* genus, *C. fatigans* in particular. However, this organism was found to be non-pathogenic to Anopheline and Aedine larval instars. The spore stage of this bacillus is affected after exposure to sunlight for 6 hours and to UV germicidal radiation for 120 minutes at a dose of 108×10^3 J/m^2, and tolerates heat treatment at 60°C for 30 minutes only. In all cases the viability and toxicity are drastically affected. It is non-toxic to *Gambusia affinis*. Small and large scale laboratory trials with *C. fatigans* larval instars produced good results. The field trials conducted in the Bombay suburbs in septic tanks with a concentration of 10^5 CFU/mL proved encouraging. Spores of this organism have a good shelf-live in a cold room (-10°C) or at room temperature as lyophilized material. This indigenously isolated *B. sphaericus* (WHO 2173) can successfully be used per se in controlling *C. fatigans* or in integrated vector control programmes.

1. INTRODUCTION

With the advent of resistant strains of human pathogens of malaria to their respective drugs, with the resurgence of resistant vector species of malaria, filariasis and encephalitis to chemical insecticides, and with the increased cost of petroleum based insecticides it was imperative to search for alternative methods of vector control. Isolation of entomopathogenic spore bearing bacilli such as *Bacillus thuringiensis* var. *israelensis* (H-14) and subsequently of a number of *Bacillus sphaericus* isolates such as Q, 1321, 1593, 2362, 2297 and 1881, which proved to be highly specific to larvae of mosquitoes in *Anopheles*, *Culex* and *Aedes* genera, paved the way for effective mosquito control programmes [1–8]. This paper reports on the data accumulated on comparative pathogenicity, environmental effects, field trials and formulations with the indigenously isolated *B. sphaericus* (ISPC-5) (WHO 2173) and establishes its feasibility as an effective mosquito control agent for locally breeding *Culex fatigans*.

2. MATERIALS AND METHODS

2.1. Isolation, identification and pathogenicity of the organism to *C. fatigans* larvae

The organism was isolated from the diseased larvae of *C. fatigans* on a nutrient agar medium. It was a terminal spore bearing gram positive bacillus. Morphological and biochemical tests revealed its identity as *B. sphaericus*. It was of the phage type IV and serotype H-26a and 26b, and showed good growth in nutrient broth and nutrient broth + 0.3% molasses. Eighteen hour old vegetative cells were non-toxic to all the larval instars of *C. fatigans*, but the endospore and spore stages were highly pathogenic. The LC_{50} value for the 2nd instar larvae was 1.7×10^4 colony forming units (CFU)/mL and the 3rd and 4th instar larvae showed a 100% mortality at 5.7×10^5 CFU/mL. It was designated ISPC-5 (WHO 2173) [3, 4].

2.2. Comparative pathogenicity

The culture was grown in a nutrient broth + 0.3% molasses medium; the endospore and spore stages were obtained by centrifugation and then lyophilized. These materials were laboratory tested against the larval instars of *Culex*, *Anopheles* and *Aedes* mosquitoes at the National Institute of Virology in Pune. Bioassays were conducted in 150 mL capacity beakers with 100 mL of tap water, using 2nd, 3rd and 4th instar larvae.

2.3. Formation of cadavers and their significance

While conducting bioassays with the 3rd and 4th instar larvae at a concentration of 10^4 to 10^5 CFU/mL, it was observed that jet black and cream coloured cadavers formed within 48 hours. These cadavers were ground individually in a tissue grinder and their CFU/mL values were determined.

3. ENVIRONMENTAL FACTORS

3.1. pH tolerance by larvae and the spores of *B. sphaericus*

pH values in the range of 4 to 9.5 were chosen for the experiments: citrate phosphate buffer with a pH of 4 to 7; boric acid borax buffer with a pH of 8 to 9; and borax–NaOH buffer with a pH of 9.5. These values were checked with an EMCO digital type EE-330 A pH meter. The experiments were done using buffered water, with tap water controls. In all cases *B. sphaericus* spores were added as treatment at 10^5 CFU/mL.

3.2. Sunlight exposure

Two litre capacity beakers with 1.5 L of tap water were used; the concentration was 5×10^6 CFU/mL. The positive control was kept at room temperature in the laboratory, another beaker was completely covered with aluminium foil, and a third was fully exposed to sunlight for 6 hours. Samples were drawn at hourly intervals and the viable counts were determined. Also, the toxicity of the samples was simultaneously determined with *Culex* larvae.

3.3. Effect of UV on spore viability and toxicity

The spore concentration was 5×10^7 CFU/mL in a 0.1M phosphate buffer at a pH of 7.0. This material was then exposed to UV germicidal tube radiation at 13 to 15 $J \cdot m^{-2} \cdot s^{-1}$. The exposure time ranged from 0 to 120 minutes. The total viable counts of the treated samples and the toxicity to the larvae were determined, and the observations were recorded.

3.4. Effect of temperature

A spore suspension of 5.3×10^7 CFU/mL was used. Heat treatment in a constant temperature water bath (40 to 90°C) was given for 30 minutes. The viability and LC_{50} values were determined with treated samples using 3rd instar larvae.

3.5. Food particle size and its influence on larval mortality

Bioassays were conducted with 3rd instar larvae using 20 to 80 mesh and 100 to 150 mesh food particles mixed with *B. sphaericus* spores. Feeding took place at hourly intervals for 6 hours and at 24 hours. The mortality rate was then recorded, along with those of the negative and positive controls.

4. LABORATORY TRIALS

4.1. Bowl experiments

Seven litre capacity bowls were used with 5 L of tap water. *C. fatigans* larvae (mixed population) were introduced (1000 in each bowl) and the final concentration of *B. sphaericus* spores was increased to 5×10^5 CFU/mL. Readings were taken after 24 hours.

4.2. Trough experiments

Fifty litre capacity plastic troughs with 35 L of tap water were used. The spore concentration was 5×10^5 CFU/mL. Five thousand larvae (mixed population) were introduced into each trough and the mortality data were recorded for 5 days.

5. SEPTIC TANK TRIALS

Field trials were undertaken in the suburbs of Bombay with the co-operation of the Pest Control Group of Greater Bombay Municipal Corporation. The final concentration was 5×10^5 CFU/mL of *B. sphaericus* spore material. The water content of the tanks was calculated and the pathogen was delivered as a fine spray. The mortality readings were recorded for 7 days. Freshly fermented and 6 year old cold storage materials were used.

6. BRIQUETTE PREPARATION AND TESTING WITH LARVAE

Briquettes were prepared by mixing Vermiculite with polyvinyl alcohol (previously irradiated at 1.67 Mrad[1]). They were mixed in proportions of 1:1:2 (PVA:H$_2$O:Vermiculite) (wt/vol./wt) and were then blended with spores (10^{10} spores/mL) in a mixer, pored into and evenly spread over the trays and lyophilized to form 1 cm^2 thick briquettes. These were tested in the laboratory for slow release, and hence toxicity. Each briquette was placed in a bowl with 4 L of water and 100 3rd instar larvae were introduced. The mortality readings were recorded daily; 100 larval stages were reintroduced every fourth day and the mortalities were determined for 45 days.

---

[1] 1 rad = 1.00×10^{-2} Gy.

7. SHELF-LIFE OF *B. sphaericus*

The spore suspension in nutrient broth (6 years old) stored in a cold room at $-10\,^\circ$C and lyophilized 6 year old endospores and spores kept at room temperature were bioassayed every quarter for 6 years to determine their toxicity to mosquito larvae and their viability.

TABLE I. COMPARATIVE PATHOGENICITY OF *B. sphaericus* STRAINS AND *B. thuringiensis* var. *israelensis* (H-14)[a]

Mosquito species	% mortality			
	ISPC-5 (WHO 2173)		1593	H-14
	Spore stage	Liberated spores		
Culex fatigans	100[b]	100[e]	100[g]	100[j]
C. fatigans (resistant)	100[c]	30[f]	–	–
C. bitaeniorhynchus	100[d]	100[f]	–	100[j]
C. tritaeniorhynchus	42[b]	45[f]	100[g]	100[j]
Anopheles stephensi	0	0	100[h]	100[k]
Aedes aegypti	0	0	49[h]	100[j]
Ae. albopictus	0	0	83.5[h]	100[j]
Ae. vitattus	–	–	100[i]	–

[a] Tests were carried out at the National Institute of Virology, Pune.
[b] 1.4×10^3 CFU/mL against 3rd instar larvae.
[c] 1.4×10^4 CFU/mL against 3rd instar larvae.
[d] 1.4×10^2 CFU/mL against 2nd instar larvae.
[e] 5.6×10^3 CFU/mL against 2nd instar larvae.
[f] 5.6×10^3 CFU/mL against 3rd instar larvae.
[g] 3.3×10^2 CFU/mL against 2nd instar larvae.
[h] 3.3×10^3 CFU/mL against 3rd instar larvae.
[i] 3.3×10^3 CFU/mL against 4th instar larvae.
[j] 2.8×10^2 CFU/mL against 3rd instar larvae.
[k] 2.8×10^3 CFU/mL against 3rd instar larvae.

TABLE II. UV SENSITIVITY DATA[a] AND LARVICIDAL ACTIVITY OF B. sphaericus[b], ISPC-5 (WHO 2173)

Exposure (min)	Dose[c] ($\times 10^3$ J)	Total viable count (CFU/mL)	% mortality[d]
0	0	4.7×10^7	100
15	11.7	2.63×10^4	100
30	27.0	8.15×10^3	93.75
60	46.8	5.36×10^3	91.25
75	58.5	2.73×10^3	43.75
90	70.2	1.36×10^3	32.50
105	94.5	1.00×10^3	15.00
120	108.0	7.00×10^2	8.75

[a] The figures presented are the average of four replicates.
[b] Stock suspension of B. sphaericus (5×10^7 CFU/mL) in a 0.1M phosphate buffer at a pH of 7.0.
[c] The dose rate of the UV lamp was 13 to 15 $J \cdot m^{-2} \cdot s^{-1}$.
[d] The control replicates without B. sphaericus showed no mortality.

TABLE III. LARVAL MORTALITY DATA[a] IN BOWL EXPERIMENTS[b] USING MIXED LARVAL POPULATIONS OF C. fatigans

	Pupal counts				
Control[c]	T_1	T_2	T_3	T_4
69	7	18	12	11
120	23	12	4	16
130	19	21	26	14

[a] Mortality was recorded after 24 hours.
[b] Each bowl contained 1000 larvae in 2nd, 3rd and 4th instars; the dose was 5×10^5 CFU/mL (spores).
[c] There was no larval mortality in the controls.

8. RESULTS

The comparative pathogenicity test results are presented in Table I. It is evident that there is a varying susceptibility among the species of the same genus. Even the 3rd instar larvae of the resistant strain of *C. fatigans* had a 100% mortality at 1.4×10^4 CFU/mL (endospores) and 5.6×10^4 CFU/mL (spores). It is evident that the endospores are more toxic than the spores. Our organism was found to be non-toxic to Anopheline and Aedine larvae, and the results obtained are in conformity with those of Yousten [9].

The cadavers that developed within 48 hours of treatment were 21 to 50% black and the rest were cream coloured. The CFU values were 3.5×10^5 to 2.93×10^7 (black) and 1.5×10^5 to 3.9×10^6 (cream coloured). The black cadavers were very fragile and help build up the pathogen.

The pH range of 4 to 6.5 was most suitable for the survival of mosquito larvae and toxicity of the pathogen is maintained, as indicated by the laboratory bioassays. Exposure to sunlight for 3 hours did not affect the toxicity and viability of the spores, but 6 hours' exposure did affect both, as proved by the bioassays and the plate counts. The spores were highly sensitive to heat and the results indicated that the LC_{50} values were 9.0×10^4 (room temperature), 1.2×10^5 (40°C), 1.05×10^5 (50°C) and 8.05×10^5 (60°C). Beyond 60°C, toxicity was completely lost.

The response of the larvae to various food particle sizes indicated that feeding on the hour for 6 hours and at 24 hours on particles of 20, 40, 60, 80, 100, 120 and

TABLE IV. LARVAL MORTALITY DATA IN TROUGH EXPERIMENTS[a] USING MIXED LARVAL POPULATIONS OF *C. fatigans*[b]

Hours after treatment	Survival counts of larvae and pupae		
	Control[c]	5×10^5 CFU/mL	5×10^4 CFU/mL
24	57 (pupae)	54 (pupae)	48 (pupae)
48	50 (pupae)	10 (1st instar)	18 (1st instar)
96	40 (pupae)	3 (1st instar)	12 (1st instar)
120	35 (adults)	0	0

[a] 5000 larvae were used per treatment.
[b] The figures presented are the average of four separate experiments.
[c] There was no larval mortality in the controls.

TABLE V. FIELD TRIALS CARRIED OUT AT THE TAGOR NAGAR CAMP, GHATKOPAR, WITH B. sphaericus (ISPC-5)[a]

Septic tank No.	Volume of H_2O (L)	B. sphaericus added (10^7 spores/mL)	Pupal count	Larval count[b] Pre-treatment/ post-treatment	Reduction in 4 days (%)
1	1000	1.00	7	1000/0	100
2	500	0.50	10	85/0	100
3	750	0.75	9	100/0	100
4	500	0.50	5	40/0	100
5	750	0.75	50	100/2	98
6	500	0.50	0	15/0	100
7	500	0.50	6	600/0	100
8	750	0.75	5	1000/0	100
9	750	0.75	25	50/0	100
10	750	0.75	4	20/2	90
11	250	0.25	0	1000/0	100
12	1000	1.00	1	520/24	95.37
13	1000	1.00	50	1000/0	100
14	250	0.25	10	50/0	100
15[c]	500	0.50	200	1000/500	50
16	500	0.50	30	200/0	100
17	750	0.75	4	30/0	100
18	750	0.75	1	20/0	100
19	750	0.75	4	15/0	100
20	500	0.50	5	100/0	100
21	750	0.75	25	35/0	100
22	250	0.25	50	100/0	100
23	750	0.75	6	200/0	100
24	500	0.50	2	200/2	99
25	500	0.50	25	400/30	92.5
26	750	0.75	25	30/0	100
27	1000	1.00	50	20/0	95

[a] Conducted with the co-operation of the Insecticide Officer, Greater Bombay Municipal Corporation.
[b] An average of two scoops.
[c] Continuous leakage was detected.

150 mesh mixed with *B. sphaericus* spores gave mortalities of 0.75% (negative control), 96% (positive control) and 93, 88, 87, 72, 57, 55 and 52% for the respective particle sizes. The results of the UV radiation experiment are given in Table II. It was found that at a dose of 11.7×10^3 J there was a reduction in viability, but no change in toxicity. However, at a dose of 108×10^3 J there was a reduction in viability by a factor of 5 and mortality was reduced 12 times. This indicates that UV protectants may have a positive role to play in field trials.

The tolerance shown by *Gambusia affinis* (mosquito fish) to a high dose of *B. sphaericus* (10^9 CFU/mL) indicates that this pathogen can successfully be used with this predator for effective mosquito control.

The results of the bowl and trough experiments were quite satisfactory; they are presented in Tables III and IV, respectively. Data on the field trial using septic tanks are given in Table V. The results are encouraging, since the co-breeding *Armigeres subalbatus* larvae were also killed.

The briquette tests in the laboratory showed that there was a slow release of spores, 8×10^7 CFU/g within 48 hours. The mortality range observed was 34 to 100%, with an average of 76%. The briquettes sank on the fifth day without disintegration and remained active for 45 days, after which there was a steady decrease in activity.

9. DISCUSSION AND CONCLUSIONS

Regarding vector control, the results obtained with different species of *Culex* larvae were quite satisfactory. Even 3rd instar larvae of the resistant *C. fatigans* strain were highly susceptible to a dose of 10^4 CFU/mL, which is most encouraging. Also, it was found that the endospore stage of this organism is more toxic than the spore stage. The non-susceptibility of the Anopheline and Aedine larvae narrows down the spectrum of activity of our pathogen. Our observations are similar to those found by other workers [3-8, 10]. In particular, development of black cadavers is a unique characteristic of our organism, and a good deal of inoculum of *B. sphaericus* was pumped into the breeding places because of its easy disintegration. The spores then persist and form the focii for future incoming larval populations. Similar observations on the persistance of spores have been made with *B. sphaericus* 2362 and 1593 against *C. pipiens* larvae [9, 11].

pH conditions ranging from 4 to 10 were ideal for our field trials, since they do not affect the toxicity of the spores. Our organism is affected by sunlight exposure, i.e. after 6 hours it lost its viability and toxicity. According to Mulligan [5] *B. sphaericus* 1593 only lost its toxicity after 5 to 6 hours' exposure, but retained its viability. The size of the food particles does help to obtain high mortality percentages, while fine particles lead to feeding competition, and hence low mortality. Our findings corroborate the findings of other workers (12, 13].

In our case, prolonged exposure (2 hours) to a UV germicidal lamp drastically reduced the viability of the spores, from 5.7×10^5 to 7.0×10^2, and the mortality percentage, from 100 to 8.75%. Burke et al. [14] showed that exposure of 1593 spores for 4 hours to a sterilizing dose of UV radiation rendered them non-viable, but their toxicity was retained. This suggests the differential response of some *B. sphaericus* isolates to UV exposure.

Feeding experiments carried out with *G. affinis* have shown that they tolerate even higher doses (10^9 CFU/mL) of our organism. This indicates that our organism can be used along with larvivorous fish in mosquito control programmes.

The bowl and trough experiments carried out in our laboratory with mixed larval populations of *C. fatigans* have provided encouraging results (see Tables III and IV). All the larval stages, except for the late 4th instar stage, pupated. This enabled us to pursue field trials under highly polluted conditions such as in the septic tanks situated in the Bombay suburbs. The data in Table V show that even the co-breeding *A. subalbatus* non-vector species was killed with *B. sphaericus*, which is advantageous [15, 16]. The slow release experiments will definitely be of help in that the pathogen source can be made available throughout the year [17, 18].

The shelf-life of our *B. sphaericus* (WHO 2173) will be of use in our mass multiplication programmes as it will be possible to run large scale field trials.

It can be concluded that our indigenous isolate ISPC-5 (WHO 2173) *B. sphaericus* is highly pathogenic to the mosquito species of the *Culex* genus. It can be grown easily and at low cost and hence can be used per se or in combination with predaceous fish. Since it has a good shelf-life, many good slow release formulations can be achieved. Therefore, this organism is most suitable for controlling *C. fatigans*, the vector of filariasis and encephalitis.

REFERENCES

[1] KELLEN, W.R., et al., *Bacillus sphaericus* Neide as a pathogen of mosquitoes, J. Invertebr. Pathol. **7** 4 (1965) 442.

[2] de BARJAC, H., et al., Serological classification of *Bacillus sphaericus* strains in relation to toxicity to mosquito larvae, Appl. Microbiol. Biotechnol. **21** (1985) 85.

[3] MENON, K.K.R., et al., Isolation, identification and toxicity of a spore-bearing *Bacillus* (ISPC-5) from diseased mosquito larvae, Indian J. Exp. Biol. **20** (1982) 312.

[4] MENON, K.K.R., et al., Toxic activity and histopathological effects of *Bacillus sphaericus* (ISPC-5) on mosquito larvae, Indian J. Exp. Biol. **20** (1982) 768.

[5] MULLIGAN, F.S., III, Laboratory and field evaluation of *Bacillus sphaericus* as a mosquito control agent, J. Econ. Entomol. **71** 5 (1978) 774.

[6] SINGER, S., *Bacillus sphaericus* for the control of mosquitoes, Biotechnol. Bioeng. **22** (1980 1355.

[7] WORLD HEALTH ORGANIZATION, Rep. WHO/VBC/80.77, VBC/BCDS/80.10, WHO, Geneva (1980).

[8] ARSHAD, A., NAIR, J.K., Efficacy of *Bacillus sphaericus* Neide against larval mosquitoes (Diptera: *Culicidae*) and midges (Diptera: *Chironomidae*) in the laboratory, Appl. Microbiol. Biotechnol. **25** (1987) 143.

[9] YOUSTEN, A.A., *Bacillus sphaericus*: Microbial factors related to its potential as a mosquito larvicide, Adv. Biotechnol. Proc. **3** (1984) 315.

[10] VANKOVA, J., Persistence and efficacy of *Bacillus sphaericus* strains 1593 and 2362 against *Culex pipiens* larvae under field conditions, Z. Angew. Entomol. **98** (1984) 185.

[11] NICOLAS, L., et al., Persistence and recycling of *Bacillus sphaericus* 2362 spores in *Culex quinquefasciatus* breeding sites in West Africa, Appl. Microbiol. Biotechnol. **25** (1987) 341.

[12] RAMOSKA, W.A., PACEY, C., Food availability and period of exposure as factors of *Bacillus sphaericus* efficacy on mosquito larvae, J. Econ. Entomol. **72** (1979) 523.

[13] RAMOSKA, W.A., PACEY, C., Influence of suspended particulates on the activity of *Bacillus thuringiensis* serotype H-14 against mosquito larvae, J. Econ. Entomol. **75** (1982) 1.

[14] BURKE, W.F., et al., Effect of UV light on spore viability and mosquito larvicidal activity of *Bacillus sphaericus* 1593, Appl. Environ. Microbiol. **46** (1983) 954.

[15] HORNBY, J., et al., Persistent spores and mosquito larvicidal activity of *Bacillus sphaericus* 1953 in well water and sewage, J. Georgia Entomol. Soc. **19** 2 (1984) 165.

[16] DES ROCHERS, B., GARCIA, R., "The efficacy of *Bacillus sphaericus* in controlling mosquitoes breeding in sewer effluent", Proc. and Papers of the 51st Annual Conference of the California Mosquito and Vector Control Association, Inc., Jan. 1983.

[17] LACEY, L.A., et al., Sustained release formulations of *Bacillus sphaericus* and *Bacillus thuringiensis* (H-14) for control of container breeding *Culex quinquefasciatus*, Mosq. News **44** 1 (1984) 26.

[18] LACEY, L.A., Production and formulation of *Bacillus sphaericus*, Mosq. News **44** 2 (1984) 153.

POSTER PRESENTATIONS

IAEA-SM-297/26P

IMPACTO DE PLAGUICIDAS EN ZONAS AGRICOLAS DE LOS LLANOS ORIENTALES (COLOMBIA)

J.A. AVELLANEDA
Instituto Nacional de los Recursos Naturales
  Renovables y del Ambiente (INDERENA),
Bogotá, Colombia

Se presenta una primera investigación realizada en la zona arrocera de los Llanos Orientales Colombianos [1] sobre el impacto de los plaguicidas utilizados en el cultivo de arroz, sobre sistemas lóticos, tomando como referencia el río Upía, canales de riego y caños adyacentes a la zona arrocera del municipio de Barranca de Upía en el Meta.

Para el estudio se realizaron muestreos en el río Upía, aguas arriba, aguas abajo y dentro de cuerpos de agua de la zona arrocera, en un total de seis estaciones de muestreo. La selección de las estaciones se hizo con base en las características del río Upía, canales de riego y canales adyacentes a la zona de los cultivos de arroz.

Para cada una de las estaciones seleccionadas, se tomaron muestras de plancton, bentos (macroinvertebrados) y peces. Simultáneamente se tomaron datos de temperatura, dureza, oxígeno disuelto, alcalinidad y pH. En mustras preservadas de agua se analizó, con métodos estandarizados por APHA (1963), nitrógeno amoniacal, nitritos, nitratos, ortofosfatos, conductividad, acidez, CO_2, DQO, cloruros y alcalinidad.

Los resultados y la discusión de este trabajo se desarrollaron con base en tres aspectos, en cada uno de los cuales se plantearon las relaciones con los temas analizados en el estudio:

1) Efecto de los factores ambientales en la amplificación de los pesticidas sobre los componentes bióticos, especialmente sobre la comunidad piscícola.
2) Caracterización de las comunidades planctónicas, bentónicas y piscícolas, con relación a la influencia y stress de pesticidas sobre sus poblaciones.
3) Modelo general de funcionamiento.

De acuerdo a los análisis de las condiciones físicas y químicas, se descartan problemas de eutrofización y las respuestas más significativas están dadas por el planctón y los efectos de los plaguicidas utilizados en el cultivo de arroz.

La alta concentración de compuestos biocidas encontrados en los sedimentos hace pensar que el desequilibrio del sistema está condicionado principalmente por la contaminación.

Se encontraron residuos de compuestos biocidas en las siguientes cantidades: 0,30 ppm para aldrín y dieldrín, 1,90 ppm para DDT y lindano, y 0,5 para metilparatión. Todos sobrepasan los rangos permitidos por la ley, los cuales son: 0,001 ppm para aldrín, 0,05 ppm para DDT, 0,005 ppm para lindano y 0,007 ppm para metilparatión [2].

REFERENCIAS

[1] DONATO, J.Ch., et al., Evaluación Preliminar del Impacto de los Plaguicidas en Sistemas Lóticos, Zona Arrocera del Municipio de Barranca de Upía (Meta, Colombia), Instituto Nacional de los Recursos Naturales, Bogotá, Colombia (1986).

[2] MINISTERIO DE SALUD, Agua y Vertimientos Líquidos, Decreto 1594/84, Ministerio de Salud, Bogotá, Colombia (1984).

IAEA-SM-297/5P

RESIDUOS, SORCION Y DESORCION DEL HERBICIDA PARAQUAT EN SUELOS TROPICALES

M.A. CONSTENLA, L.E. MORA,
E. ROJAS, E. CARAZO
Universidad de Costa Rica,
San José, Costa Rica

Se analizaron por espectrofotometría [1] residuos del herbicida paraquat en muestras de suelos tropicales de Costa Rica, dedicados al cultivo del café y donde el paraquat es aplicado desde hace alrededor de 20 años, una o dos veces por año.

El rango de valores de residuos encontrados varía entre 11,8 y 49,3 $\mu g/g$.

Las recuperaciones de residuos que se obtuvieron se encontraron en el rango de 80 a 98%.

Adicionalmente, se midieron las constantes de sorción y desorción del paraquat en diez suelos cafetaleros. El cálculo de la constante de sorción se efectuó mediante

CUADRO I. VALOR DE LAS CONSTANTES DE SORCION Y DESORCION EN AGUA Y EN DISOLUCION SATURADA EN NH$_4$Cl PARA LOS DIFERENTES SUELOS ESTUDIADOS

	Valor de constante		
Suelos cafetaleros	Sorción	Desorción en agua $\times 10^3$	Desorción en NH y Cl $\times 10^3$
Sta. María de Dota	138	3,4	1,3
Dulce Nombre de Tres Ríos	429	1,1	38
Frailes de Tarrazú	265	5,8	23
Brazil de Santana	225	1,6	26
San Roque de Grecia	291	1,9	22
La Isabel de Turrialba	216	4,6	3,2
Atirro, Turrialba	293	0,51	0,48
San Marcos, Tarrazú	271	2,5	0,61
Sarchí, Valverde Vega	434	1,2	26
Barba de Heredia	223	2,5	16

la razón de la diferencia concentraciones inicial y en equilibrio del paraquat y la concentración en equilibrio [2]. La constante de desorción será por lo tanto el inverso de esa expresión

$$K_s = \frac{C_0 - C}{C}$$

En el Cuadro I se muestran los valores de las constantes de sorción y desorción en agua y en disolución saturada en cloruro de amonio para los diferentes suelos estudiados.

Se efectuó un análisis de correlación de los valores de las constantes de sorción y desorción para cada uno de los suelos con respecto a contenido de arcilla, materia orgánica, pH (KCl), contenido de calcio y capacidad de intercambio catiónico.

Para los valores de las constantes de sorción se encontró que el único factor que estadísticamente tiene influencia es el contenido de arcilla en los suelos estudiados al nivel del 1%.

Para las constantes de desorción en agua no se dió ninguna correlación significativa.

En la desorción con NH_4Cl se encontró una alta correlación con un grado de significancia (nivel del 1%) en el siguiente orden: contenido de arcilla, capacidad de intercambio catiónico, contenido de materia orgánica y pH (KCl).

Es interesante observar que el contenido de arcilla afecta inversamente el valor de la constante de desorción, a diferencia de las otras variables. El contenido de calcio no tuvo ninguna significancia estadística al correlacionarse.

REFERENCIAS

[1] ROJAS, C.E., Estudio de residuos, degradación y comportamiento del paraquat en tres suelos cafetaleros en Costa Rica, Tesis de Grado, Universidad de Costa Rica, 1984.
[2] RILEY, D., WILKINSON, W., TUCKER, B.V., "Biological unavailability of bound paraquat residues in soil", Bound and Conjugated Pesticide Residues, American Chemical Society Symposium Series **29** (1976) 301-353.

IAEA-SM-297/6P

RESIDUOS DE LOS HERBICIDAS DIURON Y AMETRINA EN TRES SUELOS CUBANOS

M.M. HERNANDEZ-ALFONSO, A. TRAVIESO
Instituto de Investigaciones Fundamentales en Agricultura
   Tropical 'Alejandro de Humboldt' (INIFAT),
Ministerio de la Agricultura,
La Habana, Cuba

En condiciones controladas de laboratorio se estudió el efecto de la temperatura, la humedad y la concentración inicial de herbicida en la degradación de ametrina y diurón en tres suelos. El diurón resultó más persistente que la ametrina, y los procesos degradativos dependen de una cinética de primer orden [1], aunque en dos de los suelos se obtiene un mejor ajuste con el modelo hiperbólico [2]. Para una concentración inicial de 6,5 μg de ametrina por g de suelo ferralítico rojo, con

CUADRO I. PARAMETROS DE LAS RECTAS OBTENIDAS CON EL TRATAMIENTO CINETICO DE PRIMER ORDEN EN LA DEGRADACION DE LA AMETRINA

Conc. iniciales ($\mu g/g$)	Temperatura (°C)	Humedad (%)	k (d)	$T_{\frac{1}{2}}$ (d)	r (coef. corr.)
		Suelo ferralítico rojo			
2,0	30	50	$-4,72 \times 10^{-3}$	14,7	−0,96
3,5	30	50	$-4,49 \times 10^{-3}$	15,4	−0,99
6,5	30	50	$-3,73 \times 10^{-3}$	18,6	−0,97
6,5	40	50	$-7,07 \times 10^{-3}$	9,8	−0,96
6,5	30	25	$-1,65 \times 10^{-3}$	42,0	−0,88
		Suelo oscuro plástico gleysozo			
3,5	30	50	$-10,18 \times 10^{-3}$	6,8	−0,95
6,5	30	50	$-7,17 \times 10^{-3}$	9,7	−0,96

una humedad del 50% de su capacidad de retención de agua, al incrementar la temperatura de 30 a 40°C el tiempo de vida media disminuye de 18,6 a 9,8 días, mientras que a 30°C, disminuyendo la humedad de 50 al 25%, aumenta a 42 días. La descomposición del herbicida es más rápida en el suelo oscuro plástico que en el ferralítico rojo, lo cual se atribuye a la diferencia de pH entre ambos.

Un aumento de temperatura desde 30 a 50°C eleva la velocidad de degradación del diurón tres veces más en el suelo ferralítico rojo, y sólo un tercio en el ferralítico cuarcítico amarillo, siendo en ambos casos 30°C la temperatura óptima para la reacción, la cual es más rápida en el primer suelo. Un decremento de la humedad disminuye la velocidad de descomposición en ambos suelos, lo que es más evidente en el ferralítico rojo.

La energía de activación de los procesos sugiere que los microorganismos del suelo desempeñan el papel principal en la degradación de los dos compuestos [3]. En los Cuadros I y II se exponen algunos de los resultados encontrados, concluyéndose que la marcada influencia de la temperatura, la humedad y el tipo de

CUADRO II. PARAMETROS DE LAS RECTAS OBTENIDAS CON EL TRATAMIENTO CINETICO DE PRIMER ORDEN EN LA DEGRADACION DEL DIURON

Conc. iniciales (μg/g)	Temperatura (°C)	Humedad (%)	k (d)	$T_{1/2}$ (d)	r (coef. corr.)
\multicolumn{6}{c}{Suelo ferralítico rojo}					
2	30	50	$-19,9 \times 10^{-3}$	34,8	$-0,98$
4	30	50	$-22,8 \times 10^{-3}$	30,4	$-0,96$
10	30	50	$-21,7 \times 10^{-3}$	31,9	$-0,92$
10	50	50	$-66,6 \times 10^{-3}$	10,4	$-0,92$
10	30	25	$-17,5 \times 10^{-3}$	39,6	$-0,91$
\multicolumn{6}{c}{Suelo ferralítico cuarcítico amarillo}					
2	30	50	$-21,1 \times 10^{-3}$	32,9	$-0,84$
4	30	50	$-16,5 \times 10^{-3}$	42,0	$-0,84$
10	30	50	$-14,9 \times 10^{-3}$	46,5	$-0,98$
10	50	50	$-23,9 \times 10^{-3}$	29,0	$-0,97$
10	30	25	$-14,0 \times 10^{-3}$	53,3	$-0,96$

suelo en la descomposición de la ametrina y el diurón producirá una contaminación efímera de las tierras ferralíticas rojas en el verano cálido y lluvioso de Cuba; sin embargo, durante el invierno seco, la persistencia será mayor, particularmente en los suelos oscuros plásticos y ferralíticos cuarcíticos amarillos.

REFERENCIAS

[1] HILL, G.D., et al., The fate of substituted urea herbicides in agricultural soils, Agron. J. **47** (1955) 93–104.
[2] HAMAKER, J.W., "Decomposition: Quantitative aspects", Organic Chemicals in the Soil Environment, Marcel Dekker, New York (1972).
[3] MEIKLE, R.W., Measurement and prediction of picloram disappearance rates from soil, Weed Sci. **21** (1973) 549–556.

IAEA-SM-297/39P

EFFECTS OF SLOW RELEASE FORMULATIONS OF HERBICIDES ON AQUATIC VEGETATION*

M. SOERJANI
Centre for Studies of the Environment
  and Human Resources,
University of Indonesia,
Jakarta, Indonesia

1. INTRODUCTION

Aquatic vegetation is an important factor in the success of water resource management, being a primary producer that determines the high productivity of an aquatic system, e.g. fish. However, as vegetation shows abundant growth a serious imbalance may arise between plant production and aquatic fauna consumption. Consequently, measures to control excessive vegetative growth are needed.

Aquatic weed control includes manual, mechanical, biological and chemical methods. Integrated aquatic weed management has been developed following the principles of integrated pest management. However, there is a significant trend towards increasing the use of herbicides for aquatic weed control.

Some of the important aquatic weeds in tropical Asia are: water hyacinth (*Eichhornia crassipes*), molesting salvinia (*Salvinia molesta*) and florida elodea (*Hydrilla verticillata*) [1].

2. OBJECTIVES

The purpose of the study was to evaluate the efficacy and efficiency of the use of slow release formulations of terbutryn: ethylene-vinyl copolymer formulation and rubber based formulation, and to compare their efficacy with that of a commercial wettable powder (WP) formulation.

3. MATERIALS AND METHODS

The plant materials used were water hyacinth, molesting salvinia and florida elodea, and the herbicide was terbutryn. Water hyacinth and molesting salvinia were

---

* Research carried out with the support of the IAEA under Research Contract No. 3477/GS.

planted in 20 L pots filled with 10 L of tap water, approximately 1 kg of mud and additional urea fertilizer. They were treated with the two terbutryn formulations at a rate corresponding to 1, 2 and 4 kg active ingredient (a.i.)/ha. Damage to the plants was recorded daily, i.e. the percentage mortality and the LT_{90}. The days on which 90% of the plants died were recorded and water samples were taken to measure the herbicide concentration at 1, 7 and 15 days. Florida elodea were planted in a 30 L aquarium filled with 20 L of Hoagland solution (diluted four times) [2]. The plants were treated with terbutryn copolymer corresponding to 0.125, 0.25, 0.5, 1.0 and 2.0 ppm active ingredient. Damage to the plants was recorded in the same manner as above.

4. RESULTS

(a) Water hyacinth and molesting salvinia can effectively be killed with a terbutryn formulation at 1 kg a.i./ha, and florida elodea with a terbutryn copolymer equivalent to 0.125 ppm active ingredient under stagnant water conditions; under running water conditions higher doses may be required.

(b) Water hyacinth treated with a terbutryn formulation of 1.0 kg a.i./ha had an LT_{90} of more than 40 days, which is a delay of approximately 30 days compared with the LT_{90} for plants treated with the WP formulation. Similarly, molesting salvinia had an LT_{90} of 28 days and a delay of approximately 10 days compared with those plants treated with the WP formulation.

(c) The LT_{90} of water hyacinth and molesting salvinia treated with terbutryn copolymer will depend on the dose applied, and thus the amount of herbicide released. Water in pots with plants treated with 1, 2 and 4 kg a.i./ha of terbutryn formulation had herbicide concentrations of 0.05, 0.15 and 0.4 ppm, respectively. The herbicide lost (in comparison to pots without plants) is assumed to be due to the corresponding amount absorbed by the plants. Thus, the LT_{90} of plants treated with 1, 2 and 4 kg a.i./ha are 40, 36 and 24 days, respectively.

(d) Treatment of water hyacinth and molesting salvinia with a 1 kg a.i./ha rubber formulation of terbutryn had about the same effect on the plants as that of the copolymer formulation.

REFERENCES

[1] PANCHO, J.V., SOERJANI, M., Aquatic Weeds of Southeast Asia: A Systematic Account of Common Southeast Asian Aquatic Weeds, UP Los Baños, Philippines, and BIOTROP, Bogor, Indonesia (1978) 130.

[2] HOAGLAND, D.R., ARNON, D.I., The Water Culture Method for Growing Plants without Soil, University of California, Berkeley (1938) 39.

IAEA-SM-297/13P

ИЗУЧЕНИЕ ПОВЕДЕНИЯ И КОНТРОЛЬ ПЕРСИСТЕНТНЫХ ПЕСТИЦИДОВ В ПОЧВАХ СЕЛЬХОЗУГОДИЙ

М.И. ЛУНЕВ
Центральный институт агрохимического обслуживания
сельского хозяйства Госагропрома СССР,
Москва,
Союз Советских Социалистических Республик

При оценке эколого-токсикологических характеристик пестицидов важную роль играет изучение действия отдельных и совокупных факторов на процессы миграции и детоксикации их остаточных количеств в почве. Особый интерес при проведении таких исследований представляют персистентные пестициды, которые могут длительное время сохраняться в почве и создавать экологические и токсиколого-гигиенические проблемы.

В условиях лабораторного и полевых опытов изучены процессы детоксикации изомеров гексахлорциклогексана (ГХЦГ) и сим-триазиновых гербецидов. В лабораторном опыте использованы чернозем обыкновенный (гумус составляет 6,6%, pH – 6 6; емкость поглощения – 32,3 мг-экв/100 г), темно-каштановая почва (3,3%; 7,7; 21,8 мг-экв/100 г соответственно) и серозем обыкновенный (1,4%; 7,9; 13,8 мг-экв/100 г соответственно). Динамику содержания пестицидов в почвах регистрировали при трех фиксированных значениях температуры (15°, 25° и 35°С) в 7 сроков в течение 360 суток. На основании полученных данных о динамике содержания пестицидов в почве с использованием экспоненциальной модели рассчитаны значения T_{95} – периода снижения начального содержания пестицида на 95%. С помощью данных о динамике, полученных при разных температурах, рассчитаны значения энергии активации процесса E_a.

Сопоставление значений T_{95} для различных почв, приведенных в табл. I. показывает, что длительность сохранения остатков повышается с увеличением содержания гумуса в почве. Эта зависимость хорошо аппроксимируется линейным уравнением, тангенсы угла наклона для изомеров ГХЦГ и симазина составляют 63,8; 459; 22,9 и 23,2 (% · сут.)$^{-1}$ соответственно. В нативной почве персистентность изучаемых веществ коррелирует со значениями E_a: наименее персистентному симазину отвечает наименьшее значение E_a, бета-ГХЦГ – наибольшее.

По результатам изучения персистентности симазина, атразина и пропазина в сероземе и черноземе установлено, что их персистентность убывает в указанной последовательности. Атразин и пропазин более близки по поведению и параметрам деградации, чем симазин. Значения E_a для всех трех гербицидов колеблются от 55 до 100 кдж/моль, причем для чернозема они несколько ниже, чем для серозема.

ТАБЛИЦА I. ПАРАМЕТРЫ ДЕТОКСИКАЦИИ ИЗОМЕРОВ ГХЦГ И СИМАЗИНА В ПОЧВАХ РАЗЛИЧНОГО ТИПА

Параметр	Интервал температур, °C	Почва*	Альфа-ГХЦГ	Бета-ГХЦГ	Гамма-ГХЦГ	Симазин
Среднее значение T_{95}, сут	15–35	1	435	2020	670	385
		2	550	4360	690	425
		3	790	4650	785	505
E_a для различных интервалов температур, кдж/моль	15–25	1	82,4	125,7	126,3	53,4
		2	72,2	143,4	79,4	51,6
		3	80,1	121,4	39,9	47,5
	25–35	1	8,3	51,0	3,2	11,2
		2	24,7	31,1	29,1	6,7
		3	24,8	54,0	33,9	9,2

* Почвы: 1 – серозем; 2 – темно-каштановая; 3 – чернозем.

Персистентность изомеров ГХЦГ и сим-триазиновых гербицидов оценена также по результатам полевых опытов. Установлено, что T_{95} для гамма-ГХЦГ в почвах различного типа, как правило, выше, чем для альфа-изомера. Этим объясняется тот факт, что при осуществлении контроля почв сельхозугодий на содержание остаточных количеств ГХЦГ в большинстве случаев доля проб почв, содержащих гамма-изомер, и относительное содержание изомера бывает выше, чем аналогичные показатели для альфа-ГХЦГ.

Многолетний мониторинг персистентных пестицидов в почвах сельхозугодий показывает, что во многих случаях скорость их детоксикации со временем уменьшается. Это происходит, видимо, вследствие связывания остаточных количеств пестицидов компонентами почвы. Для симазина, например, в условиях северного Казахстана отмечено увеличение T_{95} в 1,3; 1,5 и 1,9 раза на второй, третий и четвертый годы соответственно после применения гербицида относительно сезона, когда препарат был применен. В случае изомеров ГХЦГ наблюдается аналогичное замедление процессов их детоксикации в почве в последующие после применения препарата годы.

Для гибкой и объективной дифференцированной оценки эколого-токсикологической опасности пестицидов, сохраняющихся в почве сельхозугодий, использован сравнительный критерий персистентности пестицидов в почве — индекс персистентности пестицидов (ИПП):

$$\text{ИПП} = T_{95} \cdot \ln(P_M / \text{ПДК}),$$

где T_{95} — персистентность пестицида в почве (средняя или для конкретных условий), месяцы;

ПДК — норматив допустимого содержания пестицидов в почве, мг/кг;

P_M — максимально рекомендуемая доза применения пестицида, кг/га д.в.

Величину $\ln(P_м/ПДК)$ можно определить как нормативно-токсикологический коэффициент (НТК). Значения НТК для развития пестицидов существенно различаются — от отрицательных до положительных, превышающих 5,00. Обычно повышенные значения НТК имеют препараты с $P_м$ порядка 6 кг/га действующего вещества и выше. Однако на величину НТК может влиять и значение норматива ПДК [1].

При использовании ИПП уровень загрязнения почвы может считаться безопасным, если значение индекса не превышает 5, умеренно опасным — если ИПП принимает значение от 5 до 20, опасным — от 20 до 60. При значениях ИПП 60 и более уровень загрязнения почв пестицидом следует рассматривать как очень опасный.

ИПК позволяет провести и общую ранжировку ксенобиотиков по степени опасности загрязнения ими почв. Для этих целей используют средние значения персистентности отдельных ксенобиотиков или тех групп, к которым они относятся. Так, пользуясь количественной оценкой устойчивости [2], ряд гербицидов можно расположить по мере убывания опасности загрязнения ими почв в последовательности (с указанием ИПК): монурон (27,7) — диурон (16,6) — симазин (15,8) — фенурон (14,6) — атразин (13,9) — прометрин (4,83) — линурон (2,76). Для всех гербицидов здесь взято НТК по санитарно-гигиеническому нормативу. Аналогичное ранжирование можно провести для любой другой группы ксенобиотиков, для которых известны средняя персистентность и НТК.

Предлагаемый индекс можно использовать для скрининга новых химических препаратов в части их эколого-токсикологических характеристик; классификации и районирования сельскохозяйственных угодий по способности их почв к самоочищению от ксенобиотиков; дифференцированного нормирования допустимого содержания ксенобиотиков в почве; выбора оптимального ассортимента средств химизации для практического использования с учетом возможных негативных экологических последствий; прогнозирования степени загрязнения почв сельхозугодий ксенобиотиками в результате применения средств химизации сельского хозяйства.

ЛИТЕРАТУРА

[1] ЛУНЕВ М.И., Сравнительный критерий персистентности ксенобиотиков в почве. Химия в сельском хозяйстве, т. 25, 2 (1987) 66–69.

[2] KHAN, S.U., Pesticides in the Soil Environment, Elsevier, Amsterdam and New York (1980) 240.

IAEA-SM-297/25P

DISTRIBUTION OF ^{14}C-TRIDEMORPH AFTER APPLICATION AS SOIL DRENCH

R.B. MOHAMAD, T.K. LIM
Department of Agronomy and Horticulture,
Faculty of Agriculture,
Universiti Pertanian Malaysia,
Serdang, Selangor, Malaysia

R.T. HAMM
Landwirtschaftliche Versuchsstation,
BASF AG,
Limburgerhof, Federal Republic of Germany

Rubber, an important agricultural crop in Malaysia, is often exposed to powdery mildew, pink disease and white, red and brown root diseases. These are caused by *Oidium hevea, Corticium salmonicolor, Rigidoporus lignosus, Ganoderma pseudoferreum* and *Phellinus noxius,* respectively. Against these pathogens, a morpholine group of fungicides called tridemorph, which inhibits ergosterol biosynthesis [1], has been found to be effective and is thus recommended for their control. Initially, it was used as a collar protectant after excision of the infected laterals [2], but later [3, 4] it was also found to be effective against white root disease when used as a soil drench, which is an easier and cheaper technique. To be able to use this technique for the control of other diseases, a study was conducted to evaluate the behaviour of tridemorph when applied to rubber plants.

One year old rubber seedlings, grown in large clay pots, were soil drenched with ^{14}C-tridemorph at 100 mg active ingredient/plant. Tridemorph in plant and soil samples, taken at various intervals, was extracted by the Bleidner extraction technique [5] and the radioactivity was determined by liquid scintillation counting.

The chemical was rapidly taken up by the plant, with the highest accumulation in the collar and main roots. It was also detected in the fine roots and stem; the leaves of the plants had the lowest accumulation. Tridemorph was observed in the soil over a period of 12 weeks. Its distribution was influenced by the soil type. In the Munchong/Prang Series, a higher concentration was detected in the lower 12 to 25 cm soil strata, whereas in the Serdang Series it was higher in the upper 0 to 12 cm soil strata. This difference in leaching capacity was attributable to the difference in soil structure.

REFERENCES

[1] De WAARD, M.A., FUCHS, A., in Fungicide Resistance in Crop Protection (DEKKER, J., GEORGOPOULOS, S.G., Eds) (1982) 87–100.

[2] LIM, T.M., Training Manual on Crop Protection and Weed Control in Rubber Plantations, Rubber Research Institute of Malaysia, Kuala Lumpur (1979) 17–27,

[3] TRAN van Canh, Lutte contre le fomes: Nouvelle méthode d'étude, Caout. Plast. 617/618 (1982).

[4] TRANS van Canh, "Use of Calixin and Sandofan F against white root disease and black stripe of *Hevea braziliensis*", Int. Rubber Conf., Rubber Research Institute of Malaysia, Kuala Lumpur (1985).

[5] HAMM, R.T., "Behaviour of Calixin in bananas", BASF Symp., La Lima, Honduras; La Ceiba, Honduras; El Carmen, Costa Rica, BASF, Limburgerhof, Federal Republic of Germany (1983).

IAEA-SM-297/17P

RADIOISOTOPIC STUDIES ON THE CONTENT AND CONVERSION RATES OF TOXIC IMPURITIES IN MALATHION FORMULATIONS

W. REIMSCHÜSSEL, J. ADAMUS
Institute of Applied Radiation Chemistry,
Technical University,
Łódź

B. ŚLEDZIŃSKI
Institute of Organic Industrial Chemistry,
Warsaw

Poland

Malathion is an insecticide used as aqueous suspension or emulsion. Malathion that is produced in Poland and distributed commercially as Sadofos is a xylene solution of malathion (30%) and emulsifier (2%) used for the preparation of 0.3% emulsion.

Pure malathion 1 is not very toxic (LD_{50} of 12 500 mg/kg for rats). Crude malathion and its formulations contain impurities which are toxic for mammals [1]. Such impurities are formed in the production stage and can develop during the storage period. Isomalathion 2, 0,0,0-trimethylthiophosphate 3 and 0,0,S-trimethylthiophosphate 4, 0,0,S-trimethyldithiophosphate 5 and 0,S,S-trimethyldithiophosphate 6 were best recognized [2, 3] as toxic impurities. The S-methyl derivatives 2, 4 and 6 are especially toxic [4] (LD_{50}: 120, 60 and 26, respectively) because their contents can increase as the result of thiono–thiolo isomerization

$$\begin{array}{c} MeO \\ \diagdown \\ P \\ \diagup \diagdown \\ MeO S \end{array} \begin{array}{c} S-CHCOOEt \\ CH_2COOEt \end{array} \longrightarrow \begin{array}{c} MeS \\ \diagdown \\ P \\ \diagup \diagdown \\ MeO O \end{array} \begin{array}{c} S-CHCOOEt \\ CH_2COOEt \end{array} \quad (1)$$

1   2

$$\begin{array}{c} MeO \\ \diagdown \\ P \\ \diagup \diagdown \\ MeO S \end{array} \begin{array}{c} XMe \end{array} \longrightarrow \begin{array}{c} MeS \\ \diagdown \\ P \\ \diagup \diagdown \\ MeO O \end{array} \begin{array}{c} XMe \end{array}$$

X = O: 3   4   (2)

X = S: 5   6   (3)

Except for compound 2, the content of the toxic impurities in malathion formulations have not been studied because the low purity of the solvents and emulsifiers prevents the use of conventional analytical methods, e.g. gas liquid chromatography.

In our studies two problems were approached. First, use of the isotope dilution method for quantitative analysis of impurities 2 to 4 in the Sadofos samples; second, determination of the decomposition rates of these impurities and the rates of the isomerization processes (Eqs (1) to (3)) during the ageing of a Sadofos sample.

We synthesized compounds 1 to 6 labelled with ^{32}P or methyl-^{14}C. Sadofos samples with a radioactive tracer added were analysed using silica gel thin layer chromatography (TLC) on Merck plates. The radioactive substances were extracted from the gel and their mass and radioactivity were measured in solutions by UV spectrophotometry and the liquid scintillation method. The detection limit of this method is about 1 ppm.

Our results for seven samples of Sadofos stored under different conditions for 2 to 7 years indicated a considerable amount of impurities 2 to 6. One of the samples contained 8.4% isomalathion and another 8.1% of compound 5. The relative contents of compounds 2 to 6 in Sadofos and in crude malathion are similar and do not correlate with the storage period.

The decomposition rates of compounds 1 to 6 and the rates of isomerization (Eqs (1) to (3)) were determined by the radiotracer kinetic method. Samples of

Sadofos or its emulsion with the tracer added were heated at 65°C and separated using the TLC method. The radioactivity of the analysed compound and its decomposition products were measured by the liquid scintillation method. The rates of decomposition were calculated from the radioactivity changes at the initial stage. The catalytical properties of the amine and water added to the Sadofos were also studied. Our results indicate that the rates of decomposition of thiolo-isomers 4 and 6 are greater than the rates of isomerization of the corresponding thiono-isomers 3 and 5. Isomalathion 2 may, however, accumulate in Sadofos under favourable conditions when its decomposition rate is lower then the isomerization rate of malathion 1. The high hydrolysis rates of all the compounds studied in the aqueous emulsion lead to the conclusion that an emulsifiable concentrate of malathion is relatively safe for humans and mammals.

REFERENCES

[1] MILES, J.W., et al., J. Agric. Food Chem. **27** (1979) 421.
[2] KANTY, J., DRYGAS, M., GAŁAZKA, S., ŚLEDZIŃSKI, B., Organika (1978) 84.
[3] PELLEGRINI, G., SANTI, R., J. Agric. Food Chem. **20** (1972) 944.
[4] ALDRIGE, W.M., et al., Arch. Toxicol. **42** (1979) 95.

IAEA-SM-297/19P

BIODEGRADATION OF ALDRIN AND GAMMA-BHC IN TROPICAL SOILS
Control of Callosobruchus maculatus F.

H.G. MORGAN
Department of Zoology,
Fourah Bay College,
University of Sierra Leone,
Freetown, Sierra Leone

This study was designed to test whether the broad spectrum plant systemic insecticides aldrin and gamma-BHC, used extensively by local farmers to control insect pests in cowpea soils, are significantly biodegraded during the period of expected usefulness and whether they afford sufficient protection against the important cowpea pod borer *Callosobruchus maculatus* F. Cowpea was grown in field exposed experimental pots containing upland and swamp cowpea growing soils

TABLE I. LEAST SQUARES REGRESSION OF 24 HOUR PROBIT MORTALITY FOR Cryptotermes havilandi OVER TIME (weeks)

Soil type	Aldrin	Gamma-BHC
Upland	Unsterilized: y = 7.46 − 0.87x Sterilized: y = 7.61 − 0.91x	Unsterilized: y = 6.39 − 0.74x Sterilized: y = 6.38 − 0.58x
Swamp	Unsterilized: y = 7.33 − 0.88x Sterilized: y = 7.58 − 0.93x	Unsterilized: y = 6.76 − 0.73x Sterilized: y = 6.77 − 0.62x

TABLE II. MEAN PER CENT SYSTEMIC EFFICACY ± SE FOR HARVESTED COWPEA SEEDS DAMAGED BY Callosobruchus maculatus AFTER 8 WEEKS' STORAGE

Insecticide	Soil	Upland	Swamp
Aldrin	Unsterilized	17 ± 3	28 ± 5
	Sterilized	23 ± 4	19 ± 4
Gamma-BHC	Unsterilized	21 ± 3	23 ± 3
	Sterilized	8 ± 2	18 ± 3

treated once at first flowering with aldrin granules (5% active ingredient) and BHC dust (3% gamma isomer) to obtain 3 ppm (wt/wt) insecticide concentrations. Biodegrading agents were killed by heat sterilization at 150°C for 2 hours.

The insectide activity in treated soils was assayed by confining nymphs of the cowpea soil termite *Cryptotermes havilandi* Sjostedt in treated soil samples taken at weekly intervals over a 5 week period. The regression equations for probit transformed 24 hour mortalities over time are given in Table I. The results of *C. maculatus* post-harvest seed damage after 8 weeks' storage, used to measure the systemic efficacy of the insecticides, are given in Table II.

Twenty-four hour mortalities were generally lower in swamp soil (20.5% humus) than in upland soil (15.8% humus), which is consistent with the expected decrease in volatilization of the insecticides as the organic matter of the soils increases. The pH of the soil, which differed by 1.2, may have marginally influenced the hydrolysis of the insecticides. The mortality curves drawn, using the regression equations of Table I, demonstrate the expected higher mortalities from sterilized soils. Analysis of variance tests for pairs of slopes show a significant difference to sterilization only for upland gamma-BHC treated soil ($P > 0.01$). Examination of Table II shows that gamma-BHC afforded significant protection against *C. maculatus* only for sterilized upland soil ($P = 0.05$).

The following conclusions can be drawn:

(1) Gamma-BHC was biodegraded in natural upland soils
(2) Soil treatments with aldrin and gamma-BHC are equally effective in protecting cowpea, grown under normal cultivation in upland and swamp soils, against *C. maculatus*
(3) Farmers should continue to use specific anti-borer insecticides.

BIBLIOGRAPHY

BURKHARDT, C.C., FAIRCHILD, M.S., J. Econ. Entomol. **60** (1967) 1602.

LAVEGLIA, J., DAHM, P.A., Ann. Rev. Entomol. **22** (1977) 483.

TAYLOR, T.A., ALUDO, J.I.S., J. Store Products Res. **10** (1974) 123.

IAEA-SM-297/29P

INFLUENCE OF PLACEMENTS ON THE DISTRIBUTION OF CONTROLLED RELEASE ^{14}C-CARBOFURAN IN WATER AND RICE PLANTS IN THAILAND*

Paiboon PRABUDDHAM*, Nuansri TAYAPUTCH**,
Bandhit ANURUK*

*Department of Soil Science,
Faculty of Agriculture,
Kasetsart University

**Division of Agricultural Toxic Substances,
Department of Agriculture,
Ministry of Agriculture and Co-operatives

Bangkok, Thailand

A field experiment was carried out at the Klong Luang Rice Experiment Station, Pathumthani Province, to investigate the influence of placements on the distribution of controlled release ^{14}C-carbofuran in water and rice plants in an acid sulphate soil in Thailand. Surface (S) and deep (D) placements and four replicates were investigated. In each replicate, nine microplots were superimposed on to a standard control plot of a fertilizer trial transplanted with 3 week old RD_{23} rice seedlings, with a spacing of 25 × 25 cm². At each side of the standard plot, two microplots (S and D placements) were systematically arranged in the second row of the guard row area. A background microplot was also included. Each microplot was composed of one open iron cylinder, with a diameter of 25 cm and a height of 2 cm, galvanized with zinc and lined with a plastic sheet. One 4 week old RD_{23} rice seedling was placed in the middle of the cylinder 1 week after the seedlings had been planted in the standard plot. One sheet (about 0.4 × 0.8 cm² or about 0.5 g) of controlled release ^{14}C-carbofuran was enclosed in a polyester bag and placed in a vertical position, either near the soil surface or at a depth of 5 cm, or about 5 cm from the centre of the cylinder. A specially made porous cup was placed 5 cm from the bag or 2.5 cm from the cylinder frame. No pesticide was added to the background microplot. Plant samples were collected 3, 5, 9 and 15 weeks after transplanting (WAT), representing plant growth at the tillering (TL), primordia initiation (PI), flowering (FL) and maturing (MAT) stages, respectively. Flooded water and

---

* Research carried out with the support of the IAEA under Research Contract No. 4116/GS.

TABLE I. INFLUENCE OF PLACEMENTS ON THE AVERAGE DISTRIBUTION OF ^{14}C-CARBOFURAN IN WATER
(% of pesticide added)

Type of water	Weeks after transplanting (WAT)					
	1	3	5	7	9	11
Surface placement (S)						
Flooded water	57.9	33.8	15.4	3.0	0.2	0.04
Soil solution	6.2	18.5	29.3	15.6	6.6	0.40
Total	64.1	52.3	44.7	18.6	6.8	0.44
Deep placement (D)						
Flooded water	12.0	51.7	22.0	8.7	8.7	0.70
Soil solution	9.0	19.5	6.5	1.4	0.9	0.20
Total	21.0	71.2	28.5	10.1	9.6	0.90

TABLE II. INFLUENCE OF PLACEMENTS ON THE AVERAGE DISTRIBUTION OF ^{14}C-CARBOFURAN IN RICE PLANTS
(% of pesticide added)

Plant parts	Physiological stages			
	TL (3 WAT)	PI (5 WAT)	FL (9 WAT)	MAT (15 WAT)
Surface placement (S)				
Culm	11.3	0.4	2.0	1.9
Leaf blade	42.9	9.6	10.1	11.9
Leaf sheath	−7.3	3.6	16.9	3.6
Grain	0	0	0	−4.5
Total	48.8	13.6	29.0	12.9
Deep placement (D)				
Culm	16.4	1.0	0.5	1.1
Leaf blade	59.7	28.5	17.4	1.0
Leaf sheath	1.3	2.8	3.7	5.7
Grain	0	0	0	−5.6
Total	77.4	32.3	21.6	2.2

a subsurface soil solution were collected at 1, 3, 5, 7, 9 and 11 WAT. The influence of the pesticide placements on the distribution of ^{14}C-carbofuran in the water and the plants is presented in Tables I and II. In Table I it can clearly be seen that about 64% of the formulation applied at the soil surface was released 1 week after transplanting (WAT), whereas only 21% was released with deep placement. The maximum release into the water of the D placement took place at 3 WAT; the level was higher than that of the maximum release of the S placement. However, the released ^{14}C-carbofuran was in the D placement, whereas in the S placement absorption was lower (Table II). The chance of environmental contamination from the D placement is, therefore, minimum. In the S placement, however, this cannot be ruled out, especially in humid tropical countries where rice is usually cultivated in the wet season and general submergence of large areas is common after heavy rains. The bioavailability of the absorbed ^{14}C-carbofuran needs further study. Experiments are strongly recommended for other formulations and pesticides.

IAEA-SM-297/22P

PESTICIDE RESIDUES IN THAILAND

Nuansri TAYAPUTCH
Division of Agricultural Toxic Substances,
Department of Agriculture,
Ministry of Agriculture and Co-operatives,
Bangkok, Thailand

Use of pesticides in Thailand has been widespread for more than two decades, resulting in a rapid increase in the importation rate. The pesticide residue levels in environmental samples, aquatic animals, birds, agricultural products, human blood and breast milk have been monitored since 1976. The data showed that pesticides, mainly organochlorine, had accumulated in most aquatic animals, both marine and fresh water, but no alarming values were revealed. Pesticide concentrations were also determined in more than 2000 samples of water, soil and sediment from agricultural areas and water sources all over Thailand. Low levels of organochlorine residues, mainly DDT, its metabolites and dieldrin, were present, as well as traces of organophosphate and carbamate residues in some areas.

Concerning pesticide residues in agricultural products, about 3000 samples of vegetables, fruits, field crops, oil crops, animal feeds, meat and eggs have been collected since 1972. Apart from very few oil and field crop samples, more than 70% of the products contained organochlorine and organophosphate residues with amounts that were lower than the Codex maximum residue limit.

Between 1980 and 1986, surveillance of organochlorine residues in breast milk and blood samples has been carried out on lactating mothers in hospitals and on farmers in agricultural areas, respectively. The results revealed a number of breast milk samples containing DDT metabolites and some which contained dieldrin, while analysis of blood serum uncovered organochlorine residues at the ppb level.

SECOND FAO/IAEA/GSF RESEARCH CO-ORDINATION MEETING
ON RESEARCH TO DEVELOP AND EVALUATE
CONTROLLED RELEASE FORMULATIONS OF PESTICIDES
TO REDUCE RESIDUES AND INCREASE EFFICACY
USING RADIOISOTOPES

held at
Gesellschaft für Strahlen- und Umweltforschung
Munich

27 November 1987

*The following pages are a Summary Report
of the Research Co-ordination Meeting that was held
immediately after and in conjunction with the Symposium*

SUMMARY REPORT

1. INTRODUCTION

The programme on Research to Develop and Evaluate Controlled Release Formulations of Pesticides to Reduce Residues and Increase Efficacy Using Radio-oisotopes is administered by the Joint FAO/IAEA Division of Isotope and Radiation Applications of Atomic Energy for Food and Agricultural Development of the International Atomic Energy Agency and is supported by funds provided by the Federal Republic of Germany through the Gesellschaft für Strahlen- und Umweltforschung, Munich. It was initiated in 1982 for a pilot phase of two years. After promising results were achieved in selecting proper formulating agents and in obtaining formulation properties that were better than those of commercial formulations, a 3 year project phase was started. In 1987, the programme was extended for a further 2 year period.

The basis for discussion were the reports of the Planning Meeting held in Vienna in November 1982 and the 1st Research Co-ordination Meeting held in Munich in October 1984. The purpose of the 2nd Research Co-ordination Meeting was to review the progress being made by the programme participants and to discuss the methodologies and results of the field tests carried out. The two main objectives of the meeting were to co-ordinate the final phase of the programme and to make recommendations for more specific research.

Controlled release is the permeation moderated transfer of an active material from a reservoir to a target surface to maintain a predetermined concentration level for a specified period of time. Over the last decade, controlled release technology has received increasing attention in the face of growing awareness that substances ranging from drugs to agricultural chemicals are often excessively toxic and sometimes ineffective when administered or applied by conventional means. The advantages of the controlled release of pesticides in contrast to conventional formulations are: less pesticide is used, pesticide is applied less frequently, the movement of pesticides is limited, and greater safety is provided to applicators and field workers, as well as for shipping.

The scientific scope and proposed programme goals were:

(1) To formulate radiolabelled pesticides using known controlled release formulations such as encapsulation, polymerization, starch granules, laminated plastics and alginate gels
(2) To study the stability and rate of release of pesticides from formulations
(3) To evaluate formulations in laboratory, greenhouse and field tests under various environmental conditions
(4) To encourage further evaluation of the most promising formulations in large field experiments.

2. IMPLEMENTATION

During the project phase, different controlled release formulations were developed and tested. The main formulating laboratories were: the University of Gent, Belgium; the University of Newcastle upon Tyne, United Kingdom; the Gesellschaft für Strahlen- und Umweltforschung, Munich, Federal Republic of Germany; and the IAEA Laboratory, Seibersdorf, Austria. The main field testing centres were: the University of Indonesia, Jakarta, Indonesia; the Regional Research Laboratory, Hyderabad, India; the Research Institute for Irrigation, Szarvas, Hungary; the Nuclear Institute for Agriculture and Biology, Faisalabad, Pakistan; and the Kasetsart University, Bangkok, Thailand. The formulating agents included artifical and natural polymers. Lignin based controlled release formulations were developed for carbofuran, diuron, propachlor, bendiocarb and 2,4-dichlorophenoxyacetic acid (2,4-D). Ethylene-vinylacetate co-polymer formulations were developed for carbofuran, endosulfan, terbutryn chlorpyrifos, parathion and monolinuron. Different latex formulations were prepared for propanil, 2,4-D, glyphosate and terbutryn. Aminoplast formulations were used for dichlobenil and propanil. Natrosol (hydroxyethylcellulose) and alginate formulations were developed for endosulfan, lindane, dieldrin, DDT, carbofuran, 2,4-D and glyphosate.

The following target systems, identified at the 1st Research Co-ordination Meeting, were used in field experiments: paddy rice, water hyacinth, hydrilla and tsetse fly. After the initial difficulties were overcome and the formulation methods optimized, all the tested materials showed advantages over the commercial formulations. Several formulation techniques can be further improved to suit areas with extreme environmental conditions. Many biological tests are still in progress and will be reported on at the final meeting.

Radiolabelled pesticides were purchased and distributed by the IAEA Seibersdorf Laboratory. The following radiolabelled pesticides are still available in limited amounts: carbofuran, chlorpyrifos-ethyl, chlorfenvinphos, 2,4-D, glyphosate, malathion, parathion, pentachlorophenol and pirimiphosmethyl.

3. SCIENTIFIC DATA

3.1 Research carried out under IAEA Research Contracts

(1) *K.G. Das (India) (RC/4581/GS)* has completed initial evaluation (in pots) of a lignin based chlorpyrifos formulation for root-dip application and a polyethylene based carbofuran formulation.

(2) *F.F. Jamil (Pakistan) (RC/3694/GS)* has shown in preliminary experiments in pots that the carbofuran formulations provided by GSF were sufficiently

promising to warrant field trials. These will be completed by about the end of 1988. According to the results obtained, experiments will be designed to evaluate the materials in cotton crops. This study will require that work be continued for one more season.

(3) *Paiboon Prabuddham (Thailand) (RC/4116/GS)* has only been involved in the programme for 2 years. Studies have just started on the persistence and distribution of ^{14}C labelled carbofuran from controlled release formulations used on acid sulphate soils, which are found in the Philippines, Vietnam, Thailand, Indonesia and some African coastal regions.

(4) *M. Soerjani (Indonesia) (RC/3477/GS)* has completed the following studies:

    (a) Biological and laboratory studies of the effectiveness and release rates of a controlled release formulation of glyphosate used on the water hyacinth

    (b) The efficacy of controlled release formulations of glyphosate, 2,4-D, dichlobenil, propanil and terbutryn

    (c) The efficacy of controlled release formulations of 2,4-D, propanil and dichlobenil used on weeds in rice cultures.

(5) *L. Szilvássy (Hungary) (RC/3795/GS)* has conducted studies of alginate and Natrosol formulations of carbofuran, 2,4-D and glyphosate, which were developed and supplied by L. Vollner of the IAEA Seibersdorf Laboratory.

It can be concluded that by using different natural polymers, and some combinations of them, the release rates of these materials in aquatic systems could be slowed down. Simple formulation procedures and low prices would allow application of formulations also in less developed countries.

3.2. Research carried out under IAEA Research Agreements

(1) *M. Bahadir and F. Korte (Federal Republic of Germany) (RA/3471/CF)* have prepared a number of controlled release formulations in thermoplastic matrices, including low density polyethylene and ethylene-vinylacetate copolymer. A number of different forms (rope, tape and sheet) were made and the dependence of release rates on the physio-chemical characteristics was evaluated. Sheet formulations of herbicides were used successfully in the cultivation of early vegetables. Rope formulations of triazine herbicides controlled aquatic weeds in the laboratory and sheet formulations of insecticides were effective against the tsetse fly. Development of controlled release formulations of pyrethroids for soil application may extend the utility of these environmentally acceptable compounds by increasing their resistance to degradation by soil microorganisms.

(2) *E.H. Schacht (Belgium) (RA/3470/CF)* has incorporated dichlobenil into aminoplast granules in the laboratory and evaluated the release rates. The rate

of release can be varied from several weeks to over 4 months by varying the size, chemical composition and content of the active ingredient of the granules. He has also prepared modified alginate formulations containing dichlobenil; these have been sent to M. Soerjani (Indonesia) and T.K. Van (United States of America) for evaluation against aquatic weeds. Propanil has also been formulated into aminoplast granules and modified alginate beads in the laboratory for evaluation of the release rates. Release may be prolonged for up to two months. These samples have been sent to M. Soerjani for evaluation.

(3) R.M. Wilkins (United Kingdom) (RA/3826/CF) has studied the release rates of carbofuran from lignin based formulations of carbofuran and has conducted studies of this formulation in rice fields using soil injection, among other application techniques.

(4) L. Vollner (FAO/IAEA) has examined the release rates and stability of alginate, Natrosol and latex formulations. The biological evaluation of several formulations has been completed in laboratory studies. Endosulfan formulations have been developed and tested in outdoor experiments. Preliminary results have been obtained in tsetse attractant traps and on tree bark. Controlled release formulations of carbofuran, 2,4-D and glyphosate have been tested against pests in rice fields in close co-operation with L. Szilvássy. Analytical testing has been carried out on carbofuran formulations prepared by the GSF laboratory to control insect pests on rice plants in Hungary.

4. RECOMMENDATIONS

The following recommendations were made by the participants of the meeting:

(1) Research on the development of controlled release formulations of pesticides for the control of the following problems should continue:

  (a) *Tsetse fly,* with endosulfan, deltamethrin and other potential insecticides
  (b) *Aquatic weeds,* with aquatic herbicides, including 2,4-D, terbutryn and glyphosate
  (c) *Insect pests of rice and cotton,* with carbofuran and other insecticides.

(2) The laboratories preparing controlled release formulations of pesticides should maintain close contact with the research co-ordinators involved in the field testing of these formulations under local environmental conditions.

(3) In spite of the relatively short period that remains before completion of the project, new participants should be allowed to join, particularly as three contract holders, from India, Egypt and Hungary, have discontinued their work. Use of controlled release formulations offers many more advantages than commercially available conventional formulations, e.g. lower doses and

lower frequency of application; safety to the handler, applicator and the environment; stabilization of the formulation; economy; and the specificity and prolonged efficacy of the pesticides. Because of the importance of this approach to pest control, it is highly desirable that new participants join the project.

(4) Since the duration of the project is not long enough to permit extensive field scale tests of the formulations that have been prepared and tested in laboratories or small plot tests, a new project on controlled release pesticide formulations should be initiated and the outline submitted to potential donors for funding. The new project should also include work on soil applied controlled release formulations.

It was concluded that the final Research Co-ordination Meeting should be held in mid-1989.

PARTICIPANTS

M. Bahadir	Federal Republic of Germany
P.C. Dreze *(Observer)*	Belgium
M. Hussain	FAO/IAEA
G. Pfister *(Observer)*	Federal Republic of Germany
Paiboon Prabuddham	Thailand
M.J. Qureshi	Pakistan
B.C. Schiffers *(Observer)*	Belgium
M. Soerjani	Indonesia
J. C. Tjell	FAO/IAEA
J.C. Vandichel	Belgium
L. Vollner *(Scientific Secretary)*	FAO/IAEA
R.M. Wilkins	United Kingdom

CHAIRMEN OF SESSIONS

Joint Session	J.C. TJELL	FAO/IAEA
Joint Session	W. KLEIN	Federal Republic of Germany
Session 1	F.P.W. WINTERINGHAM	United Kingdom
Session 2	M. HUSSAIN	FAO/IAEA
Session 3	J.C. TJELL	FAO/IAEA
Session 4	J.R. PLIMMER	United States of America
Session 5	M.A. CONSTENLA	Costa Rica

SECRETARIAT OF THE SYMPOSIUM

Scientific Secretary:	J.C. TJELL	Joint FAO/IAEA Division of Isotope and Radiation Applications of Atomic Energy for Food and Agricultural Development, IAEA, Vienna
Administrative Secretary:	M. PROSSER	Division of External Relations, IAEA, Vienna
Proceedings Editor:	P. HOWARD KITTO	Division of Publications, IAEA, Vienna
Russian Editor:	O. MELNIK	
Spanish Editor:	L. HERRERO	

LIST OF PARTICIPANTS

Al-Zenki, S.D.H.	Radiation Protection Division, Ministry of Public Health, P.O. Box 16067, Qadeseyah, . 35851 Kuwait, Kuwait
Amonkar, S.V.	Modular Laboratory, Pest Control Section, Biochemical Group, Bhabha Atomic Research Centre, Trombay, Bombay 400085, India
Antwi, L.A.K.	Institute of Aquatic Biology, P.O. Box 38, Achimota, Ghana
Avellaneda, J.A.	Instituto Nacional de los Recursos Naturales Renovables y del Ambiente (INDEREMA), Subgerencia de Medio Ambiente, Diagonal 34, No. 5–16, Bogota, Colombia
Bahadir, M.	Institut für Ökologische Chemie, Gesellschaft für Strahlen- und Umweltforschung mbH, Ingolstädter Landstrasse 1, D-8042 Neuherberg, Federal Republic of Germany
Bakir, Y.Y.F.	Radiation Protection Division, Ministry of Public Health, P.O. Box 16067, Quadeseyah, 35851 Kuwait, Kuwait
Beitz, H.C.	Institut for Pflanzenschutzforschung, DDR-1532 Kleinmachnow, German Democratic Republic
Bull, P.S.	Permanent Mission of Australia to the IAEA, Mattiellistrasse 2–4/III, A-1040 Vienna, Austria
Chen, Z.Y.	Nanjing Agricultural University, Nanjing, Jiangsu, China
Constenla, M.A.	Universidad de Costa Rica, San Jose, Costa Rica

LIST OF PARTICIPANTS

Dieter, H.H.   Institut für Wasser-, Boden-, Lufthygiene,
Bundesgesundheitsamt,
Corrensplatz 1, 1000 Berlin 33 (West)

Dohnen, P.   Institut für Toxikologie,
Gesellschaft für Strahlen- und Umweltforschung mbH,
Ingolstädter Landstrasse 1,
D-8042 Neuherberg, Federal Republic of Germany

Drèze, P.C.   Chaire de physique et chimie-physique,
Faculté des sciences agronomiques de l'Etat,
Ministère de l'éducation nationale
et de la culture française,
8, avenue de la Faculté,
B-5800 Gembloux, Belgium

Ebing, K.W.   Biologische Bundesanstalt für Land- und Forstwirtschaft,
Bundesministerium für Ernährung,
Landwirstschaft und Forsten,
Referal 313, Postfach 140270,
D-5300 Bonn 1, Federal Republic of Germany

Edwards, V.T.   Shell Research Limited, Sittingbourne Research Centre,
Sittingbourne, Kent ME10 8AG,
United Kingdom

El Zorgani, G.A.   Entomology Section, Gezira Research Station,
Agricultural Research Corporation,
P.O. Box 126, Wad Medani, Sudan

Espinosa-González, J.   Instituto de Investigación
Agropecuaria de Panamá (IDIAP),
Universidad de Panamá,
El Dorado, Panama, Panama

Gabriel, L.
(FAO)   Central Experiment Station,
Centeno, Via Arima Post Office,
Arima, Trinidad and Tobago

Gasamagera, E.   Ministère de l'agriculture, de
l'élevage et des forêts,
B.P. 621, Kigali, Minagri, Rwanda

LIST OF PARTICIPANTS

Gözek, K.	Department of Chemistry, Ankara Nuclear Research and Training Centre, Turkish Atomic Energy Authority, Beşevler, Ankara, Turkey
Hamann, H.-J.	Gesellschaft für Strahlen- und Umweltforschung mbH, Ingolstädter Landstrasse 1, D-8042 Neuherberg, Federal Republic of Germany
Hernández-Alfonso, M.M.	Instituto de Investigaciones Fundamentales en Agricultura Tropical "Alejandro de Humboldt" (INIFAT), Ministerio de la Agricultura, Calle 2, Santiago de las Vegas, Boyeros, Havana, Cuba
Hofner, E.	Gesellschaft für Strahlen- und Umweltforschung mbH, Ingolstädter Landstrasse 1, D-8042 Neuherberg, Federal Republic of Germany
Horváth, L.	Institute of Isotopes of the Hungarian Academy of Sciences, P.O. Box 77, H-1525 Budapest, Hungary
Hussain, M.	FAO/IAEA Agrochemicals and Residues Unit, IAEA Seibersdorf Laboratory, P.O. Box 100, A-1400 Vienna, Austria
Kaufmann, D.	Building 007 (Barc-West), Agriculture Research Service, United States Department of Agriculture, Beltsville, MD 20705, United States of America
Keller, E.	Landwirtschaftliche Versuchsstation, BASF AG, D-6703 Limburgerhof, Federal Republic of Germany
Kern, M.	Pesticide Residue Analysis Laboratory, Deutsche Gesellschaft für Technische Zusammenarbeit (GTZ), Postfach 5180, D-6236 Eschborn, Federal Republic of Germany
Klein, W.	Frauenhofer Institut für Umweltchemie und Ökotoxikologie, Grafschaft Hochsauerland, D-5948 Schmallenberg, Federal Republic of Germany

LIST OF PARTICIPANTS

Knuesting, E.
: Biologische Bundesanstalt für
Land- und Forstwirtschaft,
Messeweg 11-12,
D-3300 Braunschweig, Federal Republic of Germany

Korte, F.
: Institut für Ökologische Chemie,
Gesellschaft für Strahlen- und Umweltforschung mbH,
Ingolstädter Landstrasse 1,
D-8042 Neuherberg, Federal Republic of Germany

Kraus, P.
: Biologische Forschung,
Pflanzenschutzzentrum Monheim,
Bayer AG,
D-5090 Leverkusen, Federal Republic of Germany

Kreuger, J.K.
: Department of Soil Sciences,
Swedish University of Agricultural Sciences,
P.O. Box 7072,
S-75007 Uppsala, Sweden

Kühle, J.C.
: Forschungsstelle für Terrestrische
Ökotoxikologie und Bioindikation,
Edenicher Allee 146,
D-5300 Bonn 1, Federal Republic of Germany

Kutscher, R.
: FAO/IAEA Agrochemicals and Residues Unit,
IAEA Seibersdorf Laboratory,
P.O. Box 100, A-1400 Vienna, Austria

Lunev, M.I.
: Central Institute for Agrochemical Services
to Agriculture,
USSR State Agroindustrial Committee,
Orlikov per. 1/11,
Moscow 107139, USSR

Matthies, M.
: Gesellschaft für Strahlen- und Umweltforschung mbH,
Ingolstädter Landstrasse 1,
D-8042 Neuherberg, Federal Republic of Germany

May, A.
: Dirección General de Sanidad Vegetal,
Ministerio de Agricultura y Ganadería,
Apdo. 10094, San Jose, Costa Rica

LIST OF PARTICIPANTS

Mohamad, R.B.	Department of Agronomy and Horticulture, Faculty of Agriculture, Universiti Pertanian Malaysia, 43400 UPM Serdang, Selangor, Malaysia
Morgan, H.G.	Department of Zoology, Fourah Bay College, University of Sierra Leone, PMB Freetown, Sierra Leone
Müller, P.	Institut für Biogeographie, Universität des Saarlandes, D-6600 Saarbrücken, Federal Republic of Germany
Öztürk, S.	Hendek Tarim Ilagiari, Halkali Cad. 255, Sefaköy, Istanbul, Turkey
Pfister, G.	Gesellschaft für Strahlen- und Umweltforschung mbH, Ingolstädter Landstrasse 1, D-8042 Neuherberg, Federal Republic of Germany
Plimmer, J.R.	Building 007 (Barc-West), Environmental Chemistry Laboratory, Agricultural Environmental Quality Institute, Agricultural Research Service, United States Department of Agriculture, Beltsville, MD 20705, United States of America
Prabuddham, Paiboon	Department of Soil Science, Faculty of Agriculture, Kasetsart University, Bangkok 10900, Thailand
Qureshi, M.J.	Biological Chemistry Division, Nuclear Institute for Agriculture and Biology, Jhang Road, P.O. Box 128, Faisalabad, Pakistan
Reimschüssel, W.	Institute of Applied Radiation Chemistry, Technical University, Wróblewskiego 15, PL-90-924 Łódź, Poland

LIST OF PARTICIPANTS

Schiffers, B.C.
: Chaire de phytopharmacie,
Faculté des sciences agronomiques de l'Etat,
Ministère de l'éducation nationale
et de la culture française,
8, avenue de la Faculté,
B-5800 Gembloux, Belgium

Schlössler, H.
: Isocommerz,
VE Aussen- und Binnenhandelsbetrieb,
Permoserstrasse 15,
DDR-7050 Leipzig, German Democratic Republic

Sobieszcanski, J.
: Department of Biotechnology and Food Microbiology,
Academy of Agriculture,
ul. Norwida 25, PL-50–375 Wrocław, Poland

Soerjani, M.
: Centre for Studies of the Environment and
Human Resources
University of Indonesia,
Jalan Salemba 4, Jakarta 10430, Indonesia

Sun, Jinhe
: Institute of Nuclear Agricultural Sciences,
Zhejiang Agricultural University,
Hangzhou, Zhejiang, China

Tayaputch, Nuansri
: Division of Agricultural Toxic Substances,
Department of Agriculture,
Ministry of Agriculture and Co-operatives,
Bangkok 10900, Thailand

Vandichel, J.C.
: Laboratory for Organic Chemistry,
State University Gent,
Krijgslaan 281, B-9000 Gent, Belgium

Vollner, L.
: FAO/IAEA Agrochemicals and Residues Unit,
IAEA Seibersdorf Laboratory,
P.O. Box 100, A-1400 Vienna, Austria

Wallnöfer, P.
: Landesanstalt für Ernährung,
Menzingerstrasse 54,
D-8000 Munich, Federal Republic of Germany

LIST OF PARTICIPANTS

Weber, J.B.	Department of Crop Science, School of Agriculture and Life Sciences, North Carolina State University, P.O. Box 7627, Raleigh, North Carolina 27695-7627, United States of America
Wenzelburger, J.	Bayer AG, Bleerstrasse 208, D-4019 Monheim, Federal Republic of Germany
Wilkins, R.M.	Department of Agricultural and Environmental Science, University of Newcastle upon Tyne, Newcastle upon Tyne NE1 7RU, United Kingdom
Winteringham, F.P.W.	6, Knoll Court, Knoll Hill, Sneyd Park, Bristol BS9 1QX, United Kingdom

AUTHOR INDEX

Numerals refer to the first page of a paper/poster by the author concerned

Adamus, J.: 301
Amonkar, S.V.: 277
Antwi, L.A.K.: 247
Anuruk, Bandhit: 306
Artiran, F.: 157
Avellaneda, J.A.: 289
Bahadir, M.: 195
Bakhiet, T.N.: 149
Bashir, N.: 169
Borrero de Saiz, E.: 79
Brink, N.: 101
Carazo, E.: 290
Chen, B.: 89
Chen, Z.Y.: 89
Constenla, M.A.: 123, 290
Dombovári, J.: 257
Dreze, P.: 205
El Zorgani, G.A.: 149
Espinosa-González, J.: 79
Fraselle, J.: 205
Gan, Jianying: 235
Gasia, M.C.: 205
Ghods-Esphahani, A.: 113
Gözek, K.: 157
Guerra, A.: 79
Hamm, R.T.: 300
Haq, A.: 169
Hernández-Alfonso, M.M.: 292
Horváth, L.: 177
Jamil, F.F.: 169
Kling, F.: 177
Korte, F.: 195
Kraus, P.: 1
Kreuger, J.K.: 101
Kutscher, R.: 257
Lim, T.K.: 300
Lunev, M.I.: 297

Mi, C.Y.: 89
Mohamad, R.B.: 300
Mora, L.E.: 290
Morgan, H.G.: 303
Müller, P.: 11
Naqvi, S.H.M.: 169
Narayanan, V.: 277
Navarro, M.: 79
Oncsik, M.: 257
Pfister, G.: 195
Plimmer, J.R.: 61
Prabuddham, Paiboon: 306
Qureshi, M.J.: 169
Ramón, F.: 79
Rao, A.S.: 277
Reimschüssel, W.: 301
Rojas, E.: 290
Rovira, D.: 79
Schacht, E.: 267
Schiffers, B.C.: 205
Sharaf Eldin, N.: 149
Simon, L.P.: 177
Śledziński, B.: 301
Soerjani, M.: 219, 295
Suharto, Hendarsih: 185
Sun, Jinhe: 235
Taslim, Haeruddin: 185
Tayaputch, Nuansri: 306, 308
Travieso, A.: 292
Vandichel, J.C.: 267
Vollner, L.: 113, 257
Weber, J.B.: 45
Wilkins, R.M.: 185
Winteringham, F.P.W.: 29
Ye, D.C.: 89
Zhang, Yongxi: 235